Concentrated Solar Energy for Power Generation and Chemical Processing 2019

AA009278

Topical Conference at the 2019 AIChE Annual Meeting

Orlando, Florida, USA
10 - 15 November 2019

ISBN: 978-1-7138-0554-0

Printed from e-media with permission by:

Curran Associates, Inc.
57 Morehouse Lane
Red Hook, NY 12571

Some format issues inherent in the e-media version may also appear in this print version.

Printed with permission by Curran Associates, Inc. (2020)

For permission requests, please contact AIChE
at the address below.

AIChE
120 Wall Street, FL 23
New York, NY 10005-4020

Phone: (800) 242-4363
Fax: (203) 775-5177

www.aiche.org

Additional copies of this publication are available from:

Curran Associates, Inc.
57 Morehouse Lane
Red Hook, NY 12571 USA
Phone: 845-758-0400
Fax: 845-758-2633
Email: curran@proceedings.com
Web: www.proceedings.com

TABLE OF CONTENTS

576381 Experimental Investigation of $YMnO_3$ Perovskites with Sr a-Site and Al B-Site Doping for Solar Thermochemical Fuel Production

Two-step metal oxide redox cycles employ relatively simple chemistry and enable thermodynamically favorable reduction of H_2O or CO_2 to produce H_2 or CO, respectively, at lower temperatures than allowable by thermolysis. In a typical temperature-swing cycle, the metal-oxide is reduced at high temperatures and low oxygen partial pressure, followed by re-oxidation in the presence of H_2O or CO_2 at a lower temperature. Ceria is considered to be the current state-of-the-art metal oxide due to its rapid reaction kinetics and high entropy changes during oxygen exchange, the latter of which enables thermodynamically favorable oxidation even with small temperature swings and low oxidant concentrations. However, due to its large enthalpy change during oxygen exchange, it requires high temperatures (*e.g.* 1773 K) or very low oxygen activity (*e.g.* 10^{-5} atm) to achieve notable reduction extents, thereby lowering its total fuel yield per cycle.

Many recent efforts in the thermochemical community have focused on lowering the reduction temperature through the use of emerging materials such as perovskite-oxides, of the general form ABO_3, which reduce more easily and, thus, have a higher oxygen exchange capacity. $YMnO_3$ perovskites with Sr^{2+} doping ($Y_{1-x}Sr_xMnO_3$) have been shown to achieve higher reduction extents than both ceria and similarly doped $LaMnO_3$ perovskites ($La_{1-x}Sr_xMnO_3$), and may produce more fuel depending on the oxidation conditions. The Sr^{2+} concentration is directly related to the thermodynamic properties of the material. Reduction extents were shown to increase with increasing Sr^{2+} incorporation; however, this improvement comes with the requirement of larger temperature swings and oxidant concentrations to perform the oxidation reaction under thermodynamically favorable conditions. Additionally, doping of Al^{3+} in $La_{1-x}Sr_xMnO_3$ was shown to lead to a higher oxygen exchange capacity as a result of Mn-enrichment on the surface.

Herein, we present an investigation of $Y_{0.8}Sr_{0.2}Mn_{0.6}Al_{0.4}O_3$ (YSMA8264), $Y_{0.8}Sr_{0.2}Mn_{0.4}Al_{0.6}O_3$ (YSMA8246), and $Y_{0.9}Sr_{0.1}Mn_{0.6}Al_{0.4}O_3$ (YSMA9164). Assessments of the oxygen exchange capacities of these materials via temperature programmed reduction showed a significant improvement in reduction extents when compared to ceria. Equilibrium measurements were performed under conditions typifying the reduction and oxidation steps, respectively. Thermogravimetric analysis was employed to measure the nonstoichiometry of the materials at temperatures from 1173-1473 K and oxygen activity from 1.61×10^{-4}-3.23×10^{-2} atm. To assess the viability of low temperature oxidation, an isothermal relaxation scheme was used in a custom tubular reactor. The measurements were performed at temperatures from 973-1173 K and oxygen partial pressures from 1.23×10^{-20}-2.24×10^{-13} atm. During these experiments, the oxygen partial pressure was controlled by delivering a precise mixture of H_2O and H_2, with $H_2O:H_2$ ratios ranging from 1.51-72.54. Under high oxidant-to-fuel conversion conditions (*i.e.* $H_2O:H_2$ ratios approaching 1), relatively small changes in oxygen content compared to ceria were observed for each of the materials.

573522 Understanding the Oxidation Kinetics Mechanism of Ceria Based Materials in Solar Thermochemical Gas Splitting

Ceria is commonly doped with transition metal and rare earth, particularly Zr, Hf, Pr and Gd, to increase the product yield and or stabilize gain sizes during solar thermochemical water and carbon dioxide splitting. In these cycles the neat or doped ceria is heated via concentrated sunlight to over 1400°C, causing the material to reduce and release O_2; subsequently the material is exposed to steam or carbon dioxide to re-oxidize the ceria and generate the desirable H_2 or CO. The use of dopants including Hf and Zr have proven to be useful in increasing the overall cycle efficiency while Pr and Gd help to stabilize the phases. Here we use density functional theory, and kinetic Monte Carlo simulations to explore the effects of these dopants on the kinetics of re-oxidation. We find that the reactions are not diffusion limited but rather surface limited, being either sub-surface/surface O migration, or H2 formation limited, depending on the dopant. These results correspond with existing experimental results and provide an explanation for the surface limited reaction.

571837 Functional Ceramics for Solar Thermochemical Storage, Hydrogen and Liquid Fuel Production

Functional ceramics for solar thermochemical storage, hydrogen and liquid fuel production

The intermittent nature of renewable sources is one of the biggest challenges to overcome in their widespread utilization. In case of solar energy a promising method is to store it in the form of chemical energy in compounds such as hydrogen or hydrocarbons. These products are useful as storage medium, as energy carrier, as a "solar fuel" or as an intermediate for chemical commodities. Among the fuels, hydrogen is of specific interest, since the only product of hydrogen combustion is water, which in turn is the feed to produce hydrogen with solar energy.

Thermochemical cycles utilize a combination of chemical reactions driven by heat to thermochemically decompose water; therefore they are a viable option to produce hydrogen at acceptable temperature levels. The two-step cycles can achieve high efficiencies, and the materials are noncorrosive. The high temperature, therefore thermodynamically limiting step is the reduction of the oxidized form of the redox material, while the water splitting step takes place at lower temperatures as shown in Fig. 1. The most prominent examples are redox cycles applying metal/metal oxide pairs (such as ZnO/Zn), multi-valent metal oxides (e.g. Fe_3O_4/FeO, CeO_2/Ce_2O_3), or a metal oxide with oxygen non-stoichiometry (perovskites), which are capable to reversibly incorporate oxygen in their lattice.

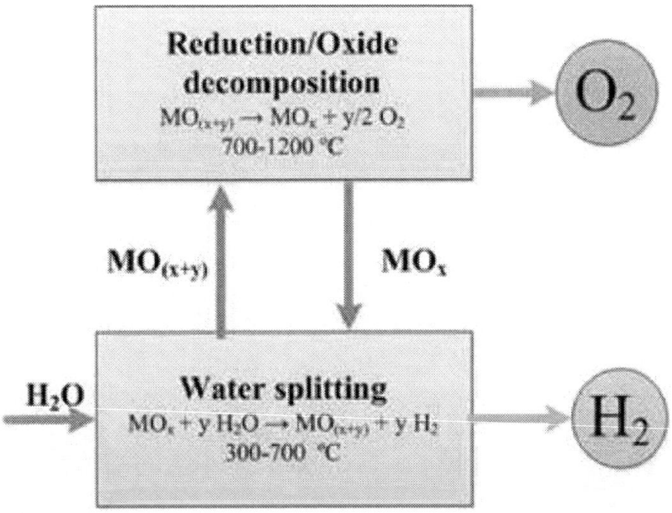

Fig. 1. Scheme of metal/metal oxide cycles for thermochemical hydrogen production.

Thermodynamic studies showed that ceria-based as well as perovskite materials are viable candidates as redox materials for thermochemical water splitting cycles. Perovskites are of special interest since they undergo a partial reduction without phase change; therefore they are more durable for long-time applications. The AMO_3 perovskite structure can be achieved by using different A and M ions, which their thermodynamic properties tuneable. In the scope of this work a library of redox materials with different compositions was established via large-scale material screening. The most promising compositions were subjected to equilibrium reduction and oxidation experiments under varying conditions to study the thermodynamic properties. Further experiments were conducted to achieve a basic understanding of interrelations between atomic structure, transport mechanisms, microstructure, reactivity and life time of the materials. The chosen functional ceramics were further developed to define and create suitable designs for solar receivers, chemical reactors and process schemes.

In parallel to material development optimizing the performance of the equipment needed to run such a process in a solar tower facility is essential. The use of particulate redox materials has been proposed to decouple the reduction reaction from the oxidation reaction, to decouple the absorption of solar radiation from the reduction reaction by using inert heat transfer particles and to allow a counter current heat exchanger design by using a combination of both particles. This concept can be implemented into any particle process and offers potentially high heat recovery rates and thus it can improve the overall system efficiencies. A detailed model to calculate the performance of the concept in consideration of temperature dependent material data and several other influencing factors is developed and validated by experiments.

Summary

The present work aims to give an overview on recent developments and the state-of-the art of high temperature fuels solar processes, in particular hydrogen production via solar thermochemical cycles. It will have a look on the most important performance parameters involved and will give an outlook on further potential and necessary developments.

572869 A Comparison of Solid-Gas Thermochemical Energy Storage Reactions and Process Configurations for CSP Applications

Concentrating solar power (CSP) integrated with thermal energy storage is considered to be a promising option to deliver cost effective and dispatchable renewable power. Thermochemical energy storage (TCES), which is the reversible conversion of solar-thermal energy to chemical energy, allows high operating temperatures, high storage density and low heat loss over long periods. Many fluid-phase and solid-gas reactions have been studied as candidates for TCES.(Pardo et al., 2014) Recently, solid-gas TCES reactions have been receiving more attention because they require less gas storage, have higher reaction temperature and easier product separation.

Today, TCES is still at an early stage of development. While considerable work has been done in reaction selection and reactor design, there are limited system-level studies for the design of integrated processes with TCES systems embedded in CSP plants. As a result, it remains unclear what is the best process configuration, reactor design, and chemistry for CSP systems with TCES. Accordingly, the goal of this work is to (1) explore process configurations for solid-gas TCES (2) evaluate the performance of different TCES strategies via optimization models, and (3) determine how the characteristics of a reversible reaction affect the selection of a TCES process configuration.

The key units of a TCES process are two reactors: R1, where the solar heat drives an endothermic reaction ( ) and R2, where the reverse exothermic reaction takes place and the reaction heat is used for power generation. To systematically develop TCES processes, we identify four key process characteristics with two options available for each: (1) R1 heat receiving (directly from solar irradiance/ indirectly from heat transfer fluid (HTF)), (2) heat transfer between R2 and power cycle (working fluid direct/ indirect contact with reactants), (3) reactor operation mode (continuous/ batch), and (4) presence of gas storage (closed-loop/ open-loop). The selection of an option for each characteristic leads to a configuration. After a preliminary screening on the possible combinations, we select the six most promising configurations for this analysis.

A TCES strategy is defined by the combination of a reversible reaction and a process configuration. We select three reactions, each representing one reaction group: (Carbonate group), (hydroxide group), and (redox group). Carbonate and hydroxide reactions can only be paired with closed-loop configurations, while redox reactions can be combined with all six configurations. Therefore, the three reactions and six configurations selected in this work lead to 12 TCES strategies in total.

We develop optimization-based process synthesis models to investigate the energetic and economic performance of these 12 strategies (Peng et al., 2017). We build models for different TCES units. Special emphasis is placed on the modeling of fixed-bed reactors, which operate in a cyclic batch mode and their first-principal model involves partial differential equations (PDEs) in time and space. Solving this first-principal model, especially for optimization, is computationally expensive, so we build surrogate models with simple algebraic equations to replace the complex first-principal models. Specifically, we first identify key input and output (e.g. heat storage and reaction conversion) variables; then generate data points using expensive numerical simulations implemented in COMSOL (which are carried out offline); generate surrogate models via data fitting; and, finally, integrate the surrogate models into the CSP-TCES process synthesis optimization model.

To design the plant under seasonal and daily variability in solar irradiance, we formulate the optimization problem as a two-stage stochastic programming model with day and night modes nested inside each scenario (Peng et al., 2018). By solving the model, we obtain the first stage design decisions (e.g. solar field area and all equipment sizes) and second stage operational decisions (e.g. stream flow rates, unit operating conditions) that yield the minimum plant levelized cost of electricity (LCOE). Other performance metrics such as energy efficiency and storage density of the optimal design are also calculated.

The analyses allow us to identify cost-efficient TCES systems. Among the 12 strategies, the hydroxide reaction + configuration with direct irradiated R1 and continuous reactor shows the best economic performance, achieving a 20% LCOE reduction over the commercial two-tank molten salt CSP plants. From this analysis, several general design rules and insights are also generated to serve as guidelines for the future development of TCES processes.

Reference

Pardo, P., Deydier, A., Anxionnaz-Minvielle, Z., Rougé, S., Cabassud, M., Cognet, P., 2014. A review on high temperature thermochemical heat energy storage. Renew. Sustain. Energy Rev. 32, 591–610. https://doi.org/10.1016/j.rser.2013.12.014

Peng, X., Root, T.W., Maravelias, C.T., 2018. Optimization-based Process Synthesis under Seasonal and Daily Variability: Application to Concentrating Solar Power. AIChE J. 0, 1–16. https://doi.org/10.1002/aic.16458

Peng, X., Root, T.W., Maravelias, C.T., 2017. Storing Solar Energy with Chemistry: The Role of Thermochemical Storage in Concentrating Solar Power. Green Chem. 2427–2438. https://doi.org/10.1039/C7GC00023E

564517 Flexible Heat Integration to Enhance Solar Energy Utilization in a Hybrid CSP Plant

Concentrated solar power (CSP) represents a potential means to harness and dispatch solar energy on a large, centralized level. However, like other renewable energy sources, CSP must deal with issues of transiency and flexible operation if it is to be competitive in the modern power grid. Due to its ability to hybridize with fossil fuels through equipment commonality and integrate thermal energy storage (TES), these issues can be mitigated. Further penetration can be achieved by employing holistic automation of CSP systems to leverage TES and hybrid operation. This work proposes a novel operation scheme, referred to here as flexible heat integration (FHI), to enhance solar energy utilization in a solar central receiver—Rankine cycle (SCR-RC) power plant hybridized with natural gas. Typically, in an SRC-RC, concentrated solar energy heats a molten salt and dispatches the salt to a series of heat exchangers in the following order: the steam superheater, steam generator, and feedwater preheater. Under variable conditions (low- and medium-grade solar activity), maintaining the high salt temperatures needed for steam superheating can in result dips in collector thermal efficiency and by extension, diminished solar energy utilization. Instead, FHI explores dispatching solar energy independently to the thermal sinks based upon solar activity rather than in sequence while having hybrid operation via natural gas firing make up for the bypassed energy sinks. For example, during low solar activity, solar energy can be more efficiently collected and stored at a lower temperature and then dispatched to the correspondingly low-temperature feedwater solar preheater while gas firing generates and superheats steam. Likewise, during medium-grade solar activity, a solar salt heated to a medium-level temperature can dispatched to the steam generator and subsequently the feedwater preheater while gas superheats steam. Under such a design, additional TES tanks are needed beyond the conventional two-tank system seen in current SCR-RC power plants. This study is conducted to detail the fundamentals of FHI and how it can be employed to further improve solar energy utilization within a hybrid SCR-RC system thereby increasing CSP penetration and reducing greenhouse emissions relative to the standard SCR-RC power plant.

581347 Chemical Engineering Innovations for a Renewable Economy

Chemical engineering evolved in the twentieth century with rise in the use of fossil resources and myriad of products derived from then. An array of these products required great innovations and made huge impact on the human race, and such examples include urea, transportation fuels, polymers, water purification chemicals, etc. Going forward as we make transition from fossil resources to renewable resources such as solar energy, wind and sustainably available biomass, we again have great opportunities for innovation that would not only make great impact but redefine chemical engineering. My talk will focus on my own experiences with challenges and opportunities presented to us as chemical engineers. I will share an overall perspective on the use of biomass as sustainable source of carbon, harvesting solar energy in conjunction with growing food without any land constraint, and creating solution processed solar cells that utilize chemical engineering principles along with knowledge from other engineering disciplines.

581350 Engineering Materials for Clinical Application

Dr. Schmidt's research has resulted in numerous patents that are licensed to companies, including AxoGen, Alafair Biosciences, Smith and Nephew, and Siluria Technologies. Comments for this panel will focus on peripheral nerve applications and one key success that is the foundation for the Avance Nerve Graft from AxoGen. In addition, along the way to uncovering strategies for nerve regeneration, Dr. Schmidt and her team also discovered a biomaterials processing approach that led to a separate application of protecting tendon's after surgery, resulting in the VersaWrap Tendon Protector from Alafair Biosciences, a start-up company in Austin, Texas.

Nerve Repair: Dr. Schmidt's technology for processing decellularized nerve tissue was licensed to AxoGen in Alachua, Florida (http://www.axogeninc.com) for the AVANCE™ nerve graft. Many tens of thousands (~100,000) of AxoGen's AVANCE™ grafts have been implanted in patients at hundreds of hospitals globally.

Post-surgical Wound Management: Using a novel in situ crystallization processing technique, Dr. Schmidt's team created mechanically robust hyaluronic acid-based hydrogel films that can be easily handled, sutured, and delivered laparoscopically. This research is the basis of Alafair Biosciences, a start-up company in Austin, TX (http://www.alafairbiosciences.com), which is run by Dr. Schmidt's PhD graduate and a neurosurgeon collaborator. Alafair's first product, VersaWrap™ Tendon Protector, has been implanted in more than 500 patients since December 2017.

579318 Gasification of Organic Solid Waste Under High-Flux Solar Irradiation: Thermodynamics, Kinetics, and Fixed-Bed Reactor Evaluation

Conventional autothermal gasification processes are driven by heat derived from combustion of a portion of feedstock. Combustion generates byproducts, which contaminate the product synthesis gas (syngas) – a mixture mainly consisting of H_2 and CO. The advantages of solar-driven gasification as a means of upgrading carbonaceous feedstocks to syngas were demonstrated in prior work in the field. Here, we present a study of solar thermal dry gasification of organic waste including thermodynamics, chemical kinetics and design, fabrication and evaluation of a 2 kW_{th} fixed-bed CO_2 gasifier. Thermodynamic analyses were conducted to explore theoretical efficiency and composition limits of the process. Chemical kinetics of CO_2 gasification were investigated by employing thermogravimetric analysis with selected types of carbonaceous feedstocks including waste wood and sewage sludge. The high-flux solar irradiation tests were performed with a windowless solar gasifier with two concentric silicon carbine cavities: the inner cavity serving as a solar receiver to absorb high-flux solar irradiation and the outer cavity serving as a reactor chamber for gasification. The gasifier tests were performed under simulated sunlight provided by a new multi-source high-flux solar simulator developed at National University of Singapore. The solar flux distribution in the reactor aperture plane was measured by a water-cooled thermogage heat flux sensor and a CCD camera. The effects of CO_2/C ratio and input solar power to the reactor on the carbon conversion, cold gas efficiency, and solar-to-fuel efficiency were experimentally explored.

570223 Analysis of the Effect of Solar Pyrolysis Parameters on the Physicochemical Properties of Biochar

Carbon materials can be produced by different methods. One of them is solar pyrolysis, which is attaining renewed interest due to simplicity to transform biomass with environmental advantages. However, pyrolysis operation parameters are decisive factors that determine the morphological and chemical properties of carbon.

This work aims to study the influence of temperature and heating rate of solar pyrolysis in the properties of carbon. Pyrolysis experiments were carried out in the 25 kW solar furnace of IER-UNAM. *Agave angustifolia* biomass was processed in a glass spherical solar reactor. Temperatures between 500 and 900°C were employed, as well as heating rates between 4 and 30°C/min. The char produced was analyzed with several techniques, which include elemental analysis (CHONS), surface area by BET and DFT models and capacitance by cyclic voltammetry.

The results of the experiments suggest that solar pyrolysis temperature and heating rate scarcely impact the carbon composition, nevertheless the structure, surface area and electrochemical response are highly affected by both. The control of the operation conditions allows to produce carbon materials that exhibit promising properties to be used as catalyst supports for polymer electrolyte membrane fuel cells and electrolyzer cells applications. Consequently, the use of concentrated solar energy represents an attractive method to produce these valuable materials.

567177 Numerical Design and Experimental Evaluation of a Direct Solar Reactor for the Catalytic Dry Reforming of Methane

Methane dry reforming is a process to produce "syngas" (H2 and CO) which is a major precursor for many chemicals and ultra-clean fuels. It is also a CO2 assisted process that leads to the conversion of CO2 to higher value products and an overall reduction of CO2 emissions from the chemical industry. For this reason, there is a good number of studies and reviews towards improving the process efficiency, designing new catalysts, and utilizing solar energy. This study is part of a bigger effort that explores novel catalysts and processes for CO2 conversion, with the goal of achieving processes in which CO2 input is greater than CO2 output at any stage, with the ultimate objective of significantly lowering Qatar's CO2 footprint. This study focuses on the utilization of solar energy to activate the Atomic Layer Deposition (ALD) based catalyst and drive the dry reforming reaction directly. First, we present the numerical design of a lab scale solar tubular reactor and second, the experimental evaluation of the reactor and dry-reforming reaction conversion efficiencies.

The thermal and optical performance of a solar tubular reactor/receiver depends upon the geometry of the receiver, optical and thermophysical properties of a receiver's material and reaction conditions. Initially, out 7kW high flux solar simulator (HFSS) is modeled using a commercial software, TracePro®, to obtain the intensity and distribution of the incident radiation flux. The tubular receiver has an internal diameter of 20 mm and an "active" length of 20 mm. To achieve the maximum possible optical efficiency and homogenous heating of the receiver a novel cavity is designed which reflects and focuses the incident flux on the tubular receiver. Multiple scenarios were investigated and optimum results are presented here.

Following the solar reactor manufacturing, an experimental setup was fabricated to study the performance of the proposed catalysts under simulated on-sun conditions. Two different catalysts have been used in a fixed bed configuration (0.2 g catalyst premixed with 2 g of inert quartz), a commercial and an ALD-based Ni/Al_2O_3. The gas mixture (CH_4, CO_2, Ar, He with 5%: 5%: 1%: balance) was fed into the reactor at weight hourly space velocities (WHSV) in the range of flow 2,000 ml/min/g to 10,000 ml/min/g. The catalytic bed was directly irradiated with a high flux solar simulator using 1-5 Xe arc lamp (6 kW each) producing three different temperature levels (550 °C, 650 °C and 850 °C). Herein we report the methane reforming conversion efficiencies and catalysts' activities. Also, the morphological, chemical, and stability changes of the catalytic particles have been studied and reported here.

571234 Economic and Dynamic Performance Solar-Driven Thermochemical Production of Hydrogen Using Ceria

Economic and dynamic performance solar-driven thermochemical production of hydrogen using ceria

Alicia Bayon and Alberto de la Calle

CSIRO Energy, PO Box P.O. Box 330, Newcastle, NSW 2300, Australia

In this work, we explore the dynamic performance of a 1 MW_{th} hydrogen production plant based on ceria thermochemical cycle. This work extended from our previous dynamic modelling recently published (de la Calle and Bayon 2019) by incorporating an economic analysis of the production of hydrogen. In this presentation, we aim to explain the main critical characteristics of the plant operation that affects the overall solar-to-fuel performance: heat recovery strategy, location, heliostat field operation and thermal inertial of chemicals and components.

The elements of the plant are shown in Figure 1. The main models are: a data source for encapsulating the weather data extraction, a heliostat field layout optimized using SolarPILOT^{TM}, a rotatory cavity receiver, two tanks for storing Ceria particles, two particle conveyors that allows Ceria to move between the reactors (labelled CeO_2 pump 1 and 2), gas valves to control the pressure in the reactors, heat exchangers and two high temperature membranes for separating the product gases. The operation of the plant begins by transporting solid particles of CeO_2 from a tank (CeO_2 tank 1) to the receiver (Figure 1) when the solar energy is available. CeO_2 is reduced using a flow of N_2 to enable a low O_2 partial pressure atmosphere. In our design, the reduced Ceria (CeO_{2-d}) is stored (in the CeO_2 tank 2) at a high temperature in order to allow a continuous flow of hydrogen. The gases coming into the receiver and oxidizer are pre-heated using the gas outlet and with heat exchangers (N_2 HEX and H_2O HEX). Heat recovery is also implemented over the ceria particles moving from the oxidizer to the receiver (CeO_2 HEX). In addition, the pressure in the reactors is controlled with purge valves. For accomplishing this study, a library of typical meteorological year was developed for ten location selected.

The plant is located in Learmonth (Western Australia), selected from our previous work as the most efficient location close to a maritime port (Bayon and de la Calle 2019) showing that a maximum production of hydrogen of 17.46 tons a year. From this plant design, we have performed a further techno-economic analysis showing that the technology could produce hydrogen at a selling price of \$18 GJ^{-1} if the total Capex is around 9 million \$ and CO_2 reduction incentives are considered. We also compared our results with previous techno-economic analysis with other forms of hydrogen like electrolysis and conventional reforming. This analysis will also provide the selling price of the hydrogen and maximum Capex and Opex to be achievable for deployment this technology with a minimum net present value of \$1 million over 30 years of operation together with a maximization of the internal rate of return.

This presentation will show how the incorporation of our economic analysis into the dynamic modelling is critical to optimize the process design conditions (e.g. particle storage inventory, maximum operational temperature, size of tanks and receivers, among others). We believe this novel approach can drive to the full economic optimization of a solar thermochemical hydrogen production plant design that can be applicable to other materials for a full comparison.

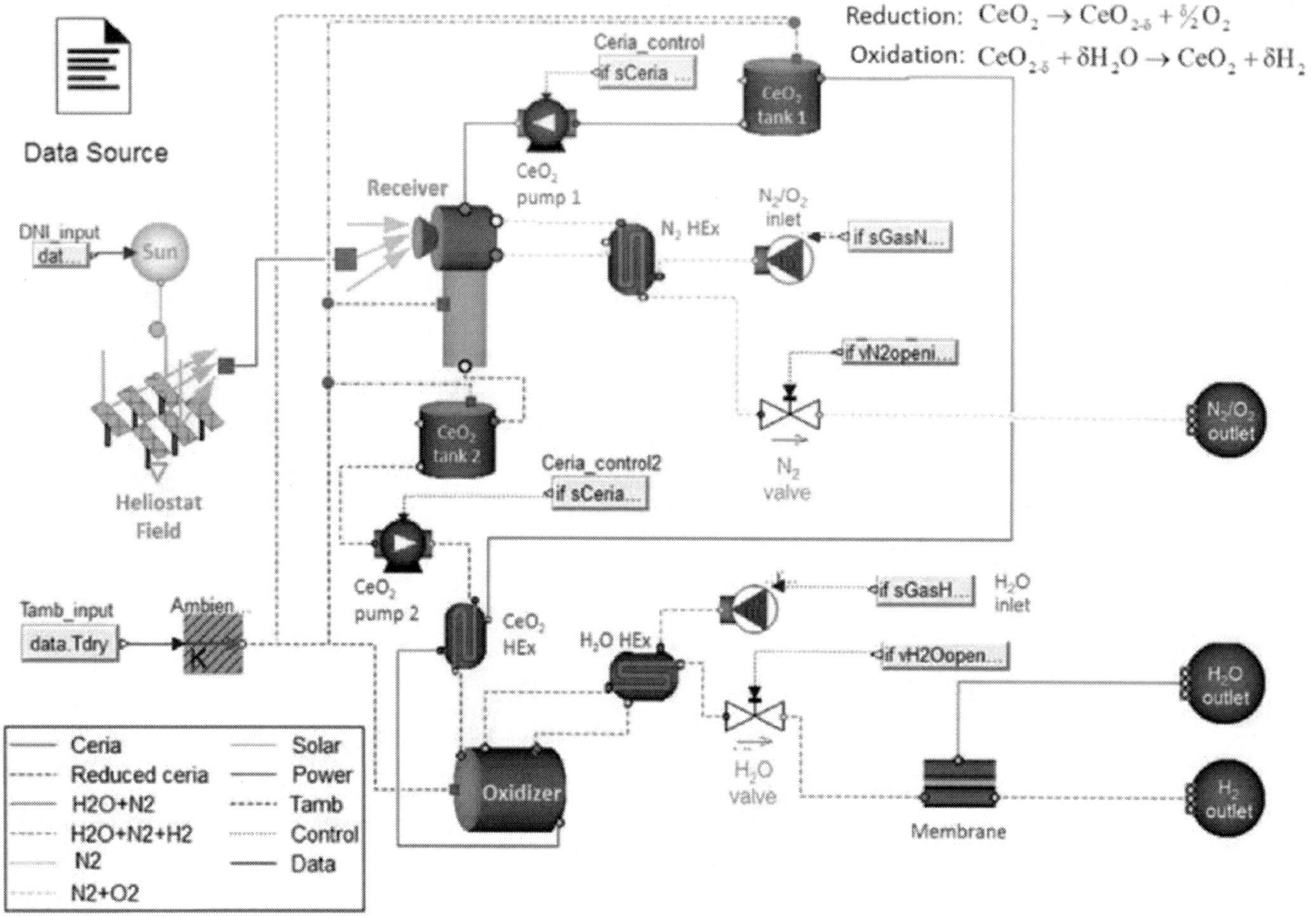

Figure 1: Modelica diagram of the solar-driven thermochemical production plant based on ceria.

Bayon, A., Calle, A. de la, 2019. Effect of Plant Location on the Annual Performance of a Hydrogen Production Plant based on CeO2 Thermochemical Cycle. AIP Conf. Proc. - SolarPaces2018 Accepted.

de la Calle, A., Bayon, A., 2019. Annual performance of a thermochemical solar syngas production plant based on non-stoichiometric CeO2. Int. J. Hydrogen Energy 44, 1409–1424. https://doi.org/10.1016/J.IJHYDENE.2018.11.076

564021 Dish-STARS™ Solar Thermochemical Production of Hydrogen

Dish-STARS™ Solar Thermochemical Production of Hydrogen

Richard F. Zheng [1], Robert S. Wegeng [2], Paul H. Humble, Johnny Saavedra-Lopez [1], Timothy G. Veldman [1]

[1] Pacific Northwest National Laboratory, Richland, WA, United States

[2] STARS Technology Corporation, Richland, WA, United States

The present paper describes the recent advances in solar thermochemical processing and reactor engineering to achieve greater than 70% solar-to-chemical energy efficiency, based on PNNL's Dish-STARS™ technology. In a Dish-STARS™ module, a Solar Thermochemical Advanced Reaction System (STARS) is coupled to a parabolic dish solar concentrator to convert solar energy into storable/useful chemical energy. The dish concentrator has been previously developed for electrical power generation. The STARS unit is a compact, process-intensive chemical reaction system based on micro- and meso-channel process technology (MMPT). Dish-STARS™ steam methane reforming has been demonstrated on-sun to produce syngas at over 70% solar-to-chemical energy efficiency, while analyses indicating a potential for values greater than 80%.

The combined Dish-STARS™ system provides a solar augment to a methane stream such as natural gas, landfill gas and biogas, increasing its chemical energy content by 20-30% while decreasing its carbon intensity. If the syngas product is further reacted to produce valuable chemical products, such as hydrogen, the reduction in carbon emissions is retained and reduced carbon intensities can be attained for the chemical products. In a co-production mode, low-carbon hydrogen or electricity plus various hydrocarbons (for example, methanol, olefins or plastics) can be produced. Currently, the Dish-STARS™ technology is under commercial development for the efficient use of concentrated solar energy in near-term applications for electrical power generation and the production of chemicals, including hydrogen (see Figure 1).

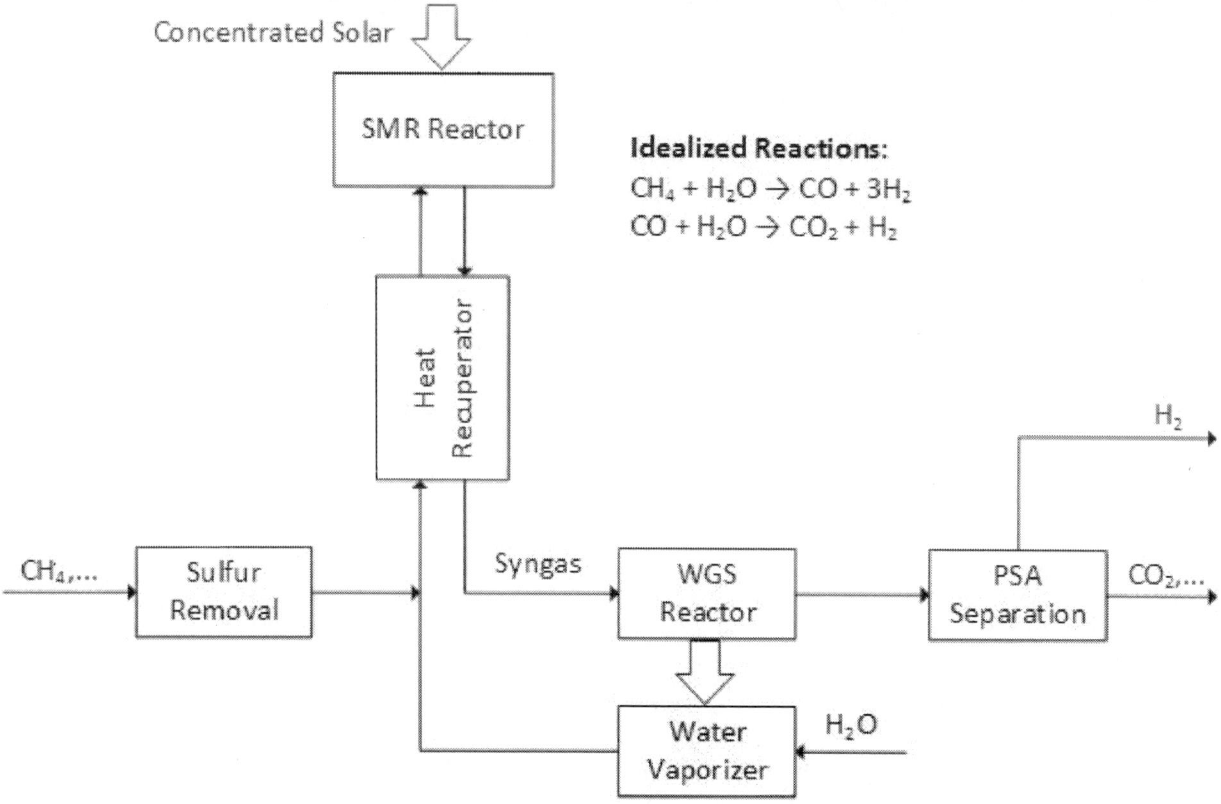

Figure 1. Level-1 block diagram of general Dish-STARS™ hydrogen production concept.

Over the course of a previous SunShot project, the STARS concept has been advanced to a technology readiness level (TRL) of 6. Reactor design was focused on improving exergetic efficiency of the SMR reactor and reducing thermomechanical stress through CFD modeling and on-sun experimental validation. A high degree of heat integration was achieved by using a number of carefully designed MMPT heat exchangers to recuperate high grade heat from the hot syngas stream. The new reactor and heat exchanger assembly was tested on-sun under high flux conditions (see Figure 2) and showed improved efficiency over previous version [1].

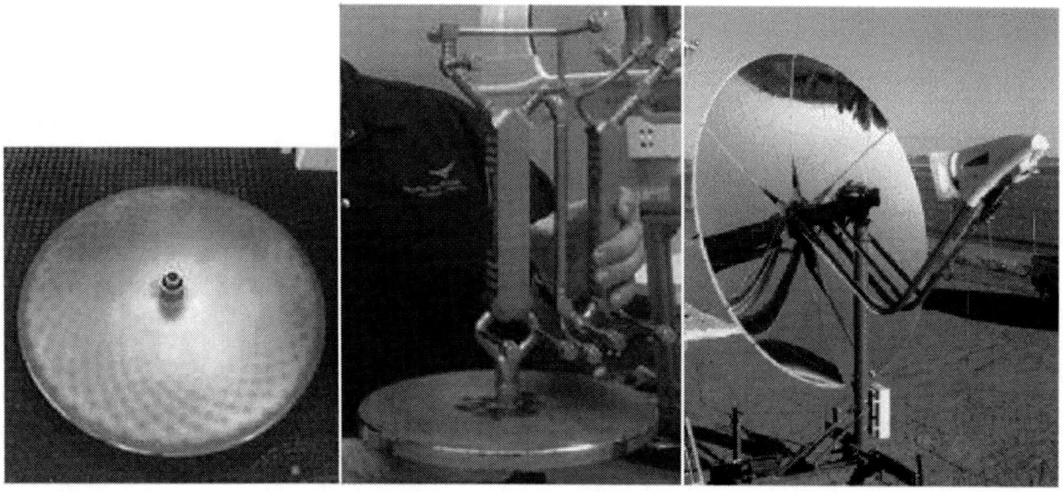

Figure 1. TRL-6 STARS reactor (left), reactor and heat exchanger assembly (middle), and the Dish-STARS™ module operated on-sun (right).

In this paper, details of the STARS reactor design to promote thermal spreading and reduce thermomechanical stress will be discussed. On-sun steam methane reforming performance data and system exergy analysis results will also be presented. Preliminary results from a manufacturing investigation will be included.

References

[1] Zheng, R.F. et al., *Energy Procedia* **69** (2015) 1192.

578447 Performance Characteristics of a Directly-Irradiated Windowless Solar Vortex Particle Receiver

We report a first-of-a-kind experimental study on the performance characteristics of a novel windowless directly-irradiated vortex-based solar particulate vessel, the Solar Expanding Vortex Receiver-Reactor (SEVR). A laboratory-scale SEVR was built and tested at a nominal capacity of 6-kWth using silicon carbide as particulate heat transfer media and a solar simulator to generate the concentrated solar radiation. A vacuum pump was used to generate a sufficiently large suction at the outlet section of the device in order to avoid particle egress through the aperture while operating the device under windowless conditions. Particle egress, flow visualisation at the aperture plane and bulk gas flow rate measurements were also carried out under iso-thermal conditions and different suction values to identify suitable conditions for carrying out hot tests. The key operating parameters, namely particle size, particle loading and gas volumetric flow rate were systematically varied to assess their influence on the thermal performance, wall cavity distribution and heat losses within the device as well as to determine the complex mechanisms controlling the behaviour of the two-phase flow in the SEVR. It was found that the improved aerodynamic of the device allows to mitigate the radiative heat losses in comparison with previous vortex-based receivers as the maximum temperature peak occurs at the opposite end to the aperture. It was also found that the the thermal performance are significantly influenced by all the parameters investigated. Particularly, the mean particle residence time, which is controlled by the particle size and gas flow rate, is a key controlling parameters influencing the maximum efficiency and temperature distribution within the device. The present investigation demonstrates for the first time that directly-irradiated vortex-based receiver technology can be efficiently operated without a window, which is the key limiting factor for scaling-up of such devices while also being a costly component subjected to damage and potential failure. The device can be employed in power generation, fuel production and high-temperature industrial processes (sensible heat, material processing) that can allow a small amount of air ingress into the receiver-reactor.

573567 Non-Thermal Equilibrium Heat Transfer Model of a Thermochemical Solar Reactor Based on Porous Media

The production of energy vectors, such as hydrogen and syngas, through thermochemical solar cycles is a promising option due to their potential high efficiencies, whose theoretical values are around 35% and 50%, respectively. To perform this process with concentrated solar energy it is necessary to develop solar reactors capable to operate with high radiative flux. An attractive reactor configuration consists on using ceramic porous media that are made or coated with the redox material. This type of reactor exhibit several advantages, which include higher volumetric yield. However, measuring temperatures inside the porous media is complicated, therefore it is necessary to develop models that allow to predict such temperatures.

In this work, a one-dimensional transient heat transfer model is proposed to evaluate the thermal behavior of a zirconium oxide foam-type porous medium exposed to concentrated solar radiation provided by the IER-UNAM Solar Furnace. The developed model considers a non-thermal equilibrium situation, in which the temperature of the porous media is different from that of the fluid. In this model an effective conductivity is used, which combines the Archie model, for the conduction contribution on the solid, and the Rosseland diffusion model, for the thermal radiation conductivity. The parameters used for the effective conductivity have been corroborated experimentally for the analyzed porous media.

The model allows calculating temperature profiles inside the porous media, which are evaluated at different values of operation parameters, such as concentrated irradiance, air flowrate, etc. The model was validated with experimental results obtained by using a 10 kW solar reactor prototype with a configuration similar to the simulated system.

Abstract 553146

ATTENUATION OF ABSORBER END LOSS IN PARABOLIC TROUGH COLLECTORS

Mohammad Abutayeh
Arkansas State University
Jonesboro, AR, USA

Kwangkook Jeong
Arkansas State University
Jonesboro, AR, USA

Anas Alazzam
Khalifa University
Abu Dhabi, UAE

Bashar El-Khasawneh
Khalifa University
Abu Dhabi, UAE

ABSTRACT

An adjustment to the conventional design of parabolic trough collectors aimed at reducing absorber end losses is suggested. The adjustment is quite simple and consists of deliberately misaligning the parabolic trough collector from its absorber in the axial direction by a predetermined amount. The structure of the parabolic trough collector remains essentially the same in the new design; however, the exposed part of the absorber will no longer be perfectly aligned with the collector due to a deliberate but calculated axial move. The cost of the new collector is expected to be equal to that of the conventional collector since the only difference between the two designs is a small misalignment. The extent of the misalignment was optimized to minimize end losses based on typical meteorological year weather data for eight locations. An Operations Research tool was employed to find the ideal level of misalignment at each location that will result in maximum incident solar radiation. A performance model of a concentrating solar power plant was then used to evaluate gains realized by the suggested misalignment at each location. The gains were modest; however, they will most likely be accomplished at almost no extra cost.

Keywords: concentrating solar power, parabolic trough collector, end loss.

INTRODUCTION

Concentrating solar power (CSP) plants are designed to focus and absorb solar heat in a solar field (SF) before forwarding it to a power block (PB) to process for electric power generation via a thermodynamic cycle. The most common method for collecting solar heat uses parabolic trough collector (PTC) assemblies to focus sunlight onto an absorber network carrying a heat transfer fluid (HTF). The HTF transfers absorbed solar heat from the SF to the PB via a heat exchanger train (HXT). A conventional PTC type CSP plant is depicted in Figure 1.

The SF of a PTC type CSP plant consists of several loops each containing a fixed number of PTC assemblies. A supply header distributes cold HTF coming from the PB among the loops and a return header collects hot HTF bound for the PB. PTC assemblies consist of numerous parabolic trough mirrors tracking the sun in an east–west rotary motion on a north–south axis as shown in Figure 2. HTF flows in an absorber pipe running in the focal line of the PTC assemblies to collect solar heat. PTC type CSP plant operation has been thoroughly described before [1]. Furthermore, proper distribution of HTF among SF loops has already been analyzed [2].

Solar heat collection in the SF is hampered by many factors like: PTC optical efficiency, mirror and absorber cleanliness, shadowing by other PTCs, loop availability, less than perfect tracking, absorber end loss, pipe heat losses, broken mirrors and glass, and other issues [3]. Improved manufacturing can raise PTC optical efficiency and tracking, while improved maintenance can increase loop availability through decreasing downtime and quickly replacing any broken mirrors or glass. Also, rapid repair of pipe and insulation ruptures will greatly minimize material and heat losses. PTC shadowing can be curtailed by increasing the row distance between SF loops; however, this will also increase the required solar collection area. The current study is focused on addressing one solvable issue: absorber end loss.

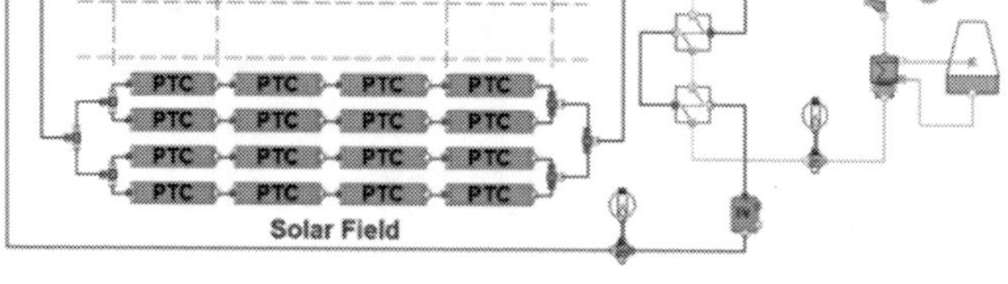

Figure 1. Conventional PTC–based CSP plant

BACKGROUND

Several researchers studied the effects of absorber end loss in PTCs on performance; however, few solutions were offered. Those few solutions involved dual axis sun tracking, which is not practical on large scale due to the linear absorber network. Even if the absorber network is somehow built to conform to the dual axis movement of the PTC, the installed cost increase would probably be intolerably significant given the large SF size. Moreover, most of those studies focused on improving the optical efficiency of PTC assemblies but overlooked the outcome on heat collection or power generation – which of course are of more significance.

A dual axis sun tracking PTC prototype was proposed [4]. The tracking is similar to conventional systems in that it tracks in an east–west rotary motion; but in addition, the entire PTC assembly was made slightly rotatable so it can align better on the north–south axis. The PTC cannot be vertically tilted, so there will always be some end loss. A small prototype was used to perform energy and exergy analysis to evaluate the advantages of this dual axis sun tracking [5]. Exergy destruction was moderately reduced and the overall exergy efficiency was increased by 3% on annual basis relative to conventional PTC assemblies.

The proposed dual axis sun tracking PTC prototype was experimentally tested with a 300 kW_{th} capacity rig [6]. The proposed PTC tracking technique reduced the solar incidence angle and end loss, consequently. Performance enhancement was felt more in the winter since the solar incidence angle is larger and therefore the end loss would have been higher in a conventional PTC assembly [7]. Also, there was little difference in performance between conventional and proposed tracking techniques in the summer due to the smaller end loss; however, this may change at different latitudes.

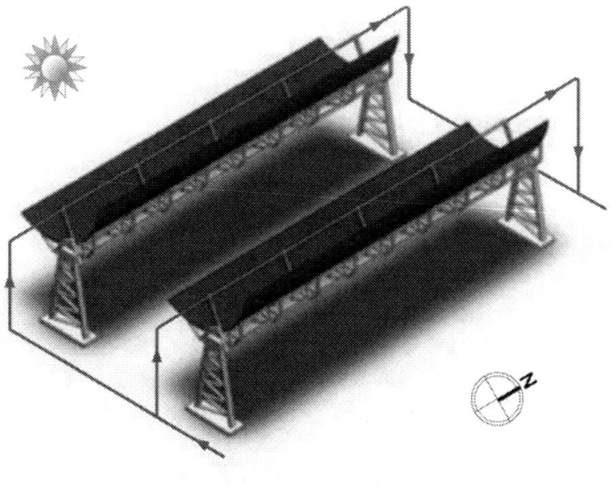

Figure 2. Conventional PTC assemblies

A similar but smaller dual axis sun tracking single PTC prototype was built and experimentally tested [8]. A computer model was also built to simulate three different tracking schemes: east–west and north–south single axis as well as dual axis. The optical efficiency and thermal performance of the PTC were notably increased when the tracking scheme was changed from single to dual axis. PTC optical efficiency was calculated based on a 24:1 concentration ratio, while PTC thermal performance was calculated based on a 200 °C HTF temperature.

A three dimensional dual axis sun tracking single PTC apparatus was built and operated using a shared sensor and software tracking mechanism [9]. Experimental data show a significant increase in heat collection compared to a fixed surface, but scaling up to industrial scale was not discussed. Another dual axis sun tracking PTC assembly was built to drive a solar thermochemical process [10]. The tracking movement was free enough to completely eliminate the end loss. A similar dual axis sun tracking PTC assembly along with its tracking control system were designed and built [11]. While PTC performance improved significantly, implementing this new PTC design in a cost effective fashion on industrial scale remains unclear.

A detailed theoretical and experimental analysis of end loss consequence on the performance of a PTC was conducted [12]. Three interesting and practically viable solutions were presented to reduce or compensate for the end loss. One of the offered solutions involve extending the exposed part of the absorber tube to capture all solar radiation reflected off the PTC. If the extension is long enough, it can totally eliminate the end loss; however, this extension will entail extra absorber cost. Another offered solution is to add a mirror at one end of the PTC structure so it can re–reflect solar radiation, that would have been otherwise lost because of the PTC end loss phenomenon, back onto the absorber. This end mirror will be attached normal to the curved mirrors of the PTC assembly which will significantly complicate mirror cleaning procedures in addition to the increased cost associated with adding a new component to the PTC structure. The end mirror will have to be located at the north end of the PTC in the northern hemisphere or at the south end of the PTC in the southern hemisphere. The last offered solution is to tilt the PTC structure to better face solar radiation. The tilt will have to be to the south in the northern hemisphere or to the north in the southern hemisphere. The ideal tilt angle that will produce the minimum average end loss is equal to the geographic latitude. The end loss will certainly be reduced by such adjustment; however, tilting is not very practical because the absorber pipe has to make turns between the PTC assemblies of the SF. This will involve extra pipe and fittings plus increased prospects for HTF leaks, which will increase installed and operating cost.

PROPOSAL

Conventional PTC assemblies track the sun by a single east–west rotary motion about a north–south axis, as shown in Figure 3. HTF flows inside an absorber pipe running in the focal line of the PTC assemblies to collect focused solar heat. This absorber pipe is exposed while running within PTC assemblies to be able to absorb reflected solar radiation, but is insulated while running between PTC assemblies to minimize heat loss from hot HTF. The exposed part of the absorber pipe is usually darkened to maximize absorptance and surrounded by a glass envelop to reduce heat loss without compromising transmittance. The exposed part of the absorber is known as a heat collection element (HCE). HCE is shown in red color while the insulated part is shown in black color in Figure 3. In conventional PTC designs, HCE is attached to PTC structures by thin supports and perfectly aligns with their mirrors. Since the sun is never fully overhead, incident and reflected sunlight will always make an acute angle with the horizontal surface of PTC mirrors known as the incident angle. Consequently, a small portion of the reflected sunlight will miss the HCE as shown in Figure 3. This sunlight reflected off the mirrors but not incident on the absorber is known as the end loss.

Altering the single axis tracking motion of the PTC to a dual axis sun tracking movement will totally eliminate the end loss; however, this is an impractical solution due to the linear absorber running in the focal line of the PTC as was mentioned before. A conceivable alternative to the dual axis tracking PTC is allow the absorber pipe to move laterally in a sliding motion in a conventional PTC design as shown in Figure 4. If the continuous lateral movement of the absorber pipe is properly controlled in such a PTC design, the end loss can be totally eliminated. This PTC design will indeed cost more than a conventional PTC design due to the nonstop sliding motion required of the absorber; therefore, it may not represent a viable option.

A practical PTC design obtained by a slight alteration of the conventional scheme is proposed. The adjustment is realized by deliberately misaligning the PTC from its absorber in the lateral direction by a fixed amount as shown in Figure 5. The size of the lateral misalignment is site specific and must be optimized for maximum incident solar radiation. The structure of the PTC and its tracking motion remain principally the same; however, the HCE will no longer perfectly align with PTC mirrors due to this lateral move. With reference to a fixed PTC, the lateral HCE shift will have to be to the north in the northern hemisphere or to the south in the southern hemisphere. The incremental cost increase associated with the new design is expected to be trivial since the only revision is a small misalignment. No extra PTC parts are required and no special tracking provisions are needed. change

Figure 3. Conventional single axis sun tracking PTC

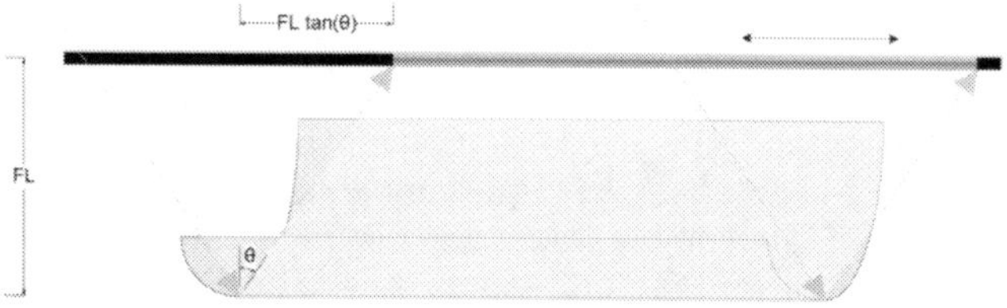

Figure 4 Optimized PTC with a sliding absorber

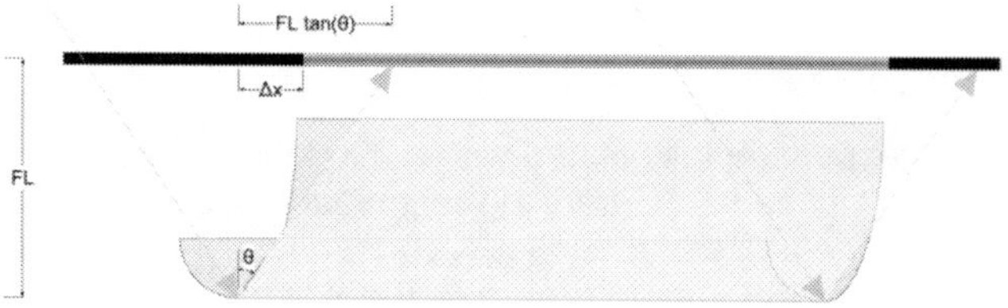

Figure 5. Proposed single axis sun tracking PTC

CALCULATION

PTC end loss calculation will be detailed next. Typical metrological year (TMY) weather data for eight different locations around the world will then be used to find the optimum PTC–absorber misalignment at each location that will minimize the impact of end loss based on the presented end loss calculation. A performance model of a 50 MW_e net rated capacity PTC type CSP plant will be run in three different PTC–absorber misalignment modes at each location: a standard mode with no misalignment as shown in Figure 3, an idealized mode with a constantly moving HCE as shown in Figure 4, and a recommended mode with the optimized PTC–absorber misalignment as shown in Figure 5. Model results for the three different PTC–absorber misalignment modes at each location will then be compared to assess the outcome of the proposed end loss attenuation scheme on solar heat collection, electric power generation, and CSP plant revenue.

The ensuing equations were disclosed in earlier work concerning CSP plant simulation [1] and are taken from research findings done at Sandia National Laboratories [13] [14]. They are used to compute solar time to then compute solar angles. Next, attenuation of solar radiation is then estimated to quantify captured solar heat.

TC is a time correction term needed to adjust regular time to solar time. TC is a function of time zone, longitude angle, and day of year and is estimated by

$$TC = 9.87 \cdot sin(2x) - 7.53 \cdot cos(x) - 1.5 \cdot sin(x) - 4 \cdot (15\,TZ - l)$$
$$x = 2\pi \cdot (Day - 81)/365 \tag{1}$$

SD is the solar day which is also needed to adjust regular time to solar time and is given by

$$SD = \begin{cases} Day, & 0 \leq Hour + \dfrac{TC}{60} \leq 24 \\ Day - 1, & Hour + \dfrac{TC}{60} < 0\ AND\ Day > 1 \\ Day + 364, & Hour + \dfrac{TC}{60} < 0\ AND\ Day \leq 1 \\ Day + 1, & Hour + \dfrac{TC}{60} > 24\ AND\ Day < 365 \\ 1, & Hour + \dfrac{TC}{60} > 24\ AND\ Day \geq 365 \end{cases} \tag{2}$$

SH is the solar hour used later to calculate several solar angles and is given by

$$SH = \begin{cases} Hour + \dfrac{TC}{60}, & 0 \leq Hour + \dfrac{TC}{60} \leq 24 \\ 24 + Hour + \dfrac{TC}{60}, & Hour + \dfrac{TC}{60} < 0 \\ Hour + \dfrac{TC}{60} - 24, & Hour + \dfrac{TC}{60} > 24 \end{cases} \tag{3}$$

Next, declination, hour, altitude, azimuth, and altitude transverse solar angles are given by

$$\delta = \sin^{-1}\left[0.39795 \cdot \cos\left(0.98563 \cdot \frac{\pi}{180} \cdot [SD - 173]\right)\right] \tag{4}$$

$$HA = \frac{\pi}{180} \cdot 15 \cdot (SH - 12) \tag{5}$$

$$\alpha = \sin^{-1}[\sin(\delta) \cdot \sin(\lambda)] + \cos(\delta) \cdot \cos(HA) \cdot \cos(\lambda) \tag{6}$$

$$a = \frac{\sin(\alpha)\ \cdot\ \sin(\lambda) - \sin(\delta)}{\cos(\alpha)\ \cdot\ \cos(\lambda)} \tag{7}$$

$$AT = \begin{cases} \alpha, & |a - \pi| < 1 \\ \tan^{-1}\left(\dfrac{\tan(\alpha)}{\left|\cos\left(\frac{\pi}{2} + a\right)\right|}\right), & |a - \pi| \geq 1 \end{cases} \tag{8}$$

SA is the shadow argument needed to evaluate shadow effects on incident solar radiation. SA is a function of PTC row distance and aperture width plus the altitude transverse solar angle

$$SA = \frac{RD}{AW} \cdot \cos\left(\frac{\pi}{2} - AT\right) \tag{9}$$

Shadow efficiency is a multiplier used in adjusting direct incident solar radiation to account for PTC shadow eclipsing the SF around sunrise and sunset

$$\eta_{Shadow} = \begin{cases} 1, & SA \geq 1 \\ 0, & SA < 0 \\ SA, & 0 \leq SA < 1 \end{cases} \tag{10}$$

IA is the incident angle defined as the angle between solar beams and the line normal to the PTC aperture. It is constantly changing and can be calculated by

$$IA = \cos^{-1}\left\{\sqrt{1 - [\cos(\alpha) \cdot \cos(a)]^2}\right\} \tag{11}$$

The above IA formula is appropriate at 0° PTC tilt angle (horizontally levelled PTC) and 0° orientation angle (north–south positioned PTC axis). Otherwise, tilt and orientation angles should be considered in the above IA calculation as was described in previous work [1].

IAM is the incident angle modifier used to adjust incident solar radiation to account for direct and indirect losses due to incident angle. IAM is a multiplier that can be estimated by the following correlation

$$IAM = \cos(IA) - 0.03008 \cdot IA - 0.09389 \cdot IA^2 \tag{12}$$

End loss efficiency is a multiplier used for adjusting incident solar radiation to account for end losses, solar radiation incident on PTC mirrors but not landing on PTC absorber or HCE. End loss efficiency is given by

$$\eta_{End\ Loss} = 1 - \frac{|FL \cdot \tan(IA)\ -\ \Delta x|}{L_{PTC}} \tag{13}$$

The term *I FL· tan (IA) – Δx I* represents the extent of HCE missed by the concentrated and reflected incident solar radiation as illustrated by Figure 3, Figure 4, and Figure 5. The quantity *Δx* represents the extent of the PTC–absorber lateral misalignment; thus, there are three possibilities for its calculation: *Δx = 0* in a standard PTC design as shown in Figure 3 signifying no misalignment, *Δx = FL· tan (IA)* in an ideal PTC design with a sliding absorber shown in Figure 4 signifying no end loss due to continual tracking, or *Δx = B* in the proposed PTC design shown in Figure 5 signifying a fixed misalignment.

Accordingly, the quantity Δx can be expressed by

$$\Delta x = \begin{cases} 0, & PTC\ Misalignment = Normal \\ B, & PTC\ Misalignment = Fixed \\ FL\tan(IA), & PTC\ Misalignment = Tracking \end{cases} \quad (14)$$

The constant B in the above equation is the proposed PTC–absorber misalignment. Its value will be different at different locations as adjusted by an operations research tool to maximize incident solar radiation.

Absolute efficiency is an overall multiplier used to compute the incident part of available solar radiation. In the context of a PTC based CSP plant, it combines the effects of optical efficiency, mirror cleanliness, shadowing, incident angle, percent of PTC assemblies in service, and end loss [15]. Absolute efficiency, $\eta_{Absolute}$, is given by

$$\eta_{Absolute} = \eta_{Optical} \cdot \eta_{Shadow} \cdot \eta_{End\ Loss} \cdot C_{Mirror} \cdot IAM \quad (15)$$

Direct normal irradiance (DNI) is beam solar radiation available for exploitation, while direct incident irradiance (DII) is the slice of DNI that is incident on PTC absorbers. DII is given by

$$DII = \eta_{Absolute} \cdot DNI \quad (16)$$

Solar times and angles in addition to DNI attenuation efficiencies were calculated using the above equations at eight different locations spatially spread over the globe. The calculations were performed in one hour increments for a whole year to match TMY data format. TMY weather data at those eight different locations were obtained from the National Renewable Energy Laboratory (NREL) web site. The same standard *EuroTrough* design was used for PTC assemblies at all locations; furthermore, the same generic values of optical efficiency, mirror cleanliness, and PTC availability were used at all locations.

The above equations were coded in *Microsoft Excel*. *Solver* is a *Microsoft Excel* add–in that can be used to perform operations research type analysis on large data sets. *Solver* was used to find the optimum PTC–absorber misalignment extent, *B*, at each site that will maximize the sum of incident solar radiation over the TMY period of each location: $\Sigma\ (DII \cdot \Delta t)_{Maximum} = f\ (B_{Optimum})$.

A performance model of a 50 MW_e net rated capacity PTC type CSP plant with a properly sized thermal energy storage (TES) system will be run in the aforementioned PTC–absorber misalignment modes. Model results of captured solar heat and power generation for the three different PTC–absorber misalignment modes at each location will then be compared to assess the outcome of the proposed end loss attenuation scheme on solar heat.

RESULTS

End loss is constantly changing depending on solar times and angles. Figure 6 illustrates four daily variations of end loss at an undisclosed location in the northern hemisphere, hence the hidden y–axis values. Since solar radiation is experienced between sunrise and sunset, end loss and other solar calculations before sunrise or after sunset are irrelevant. End losses are least in the summer and most in the winter during the daytime operation.

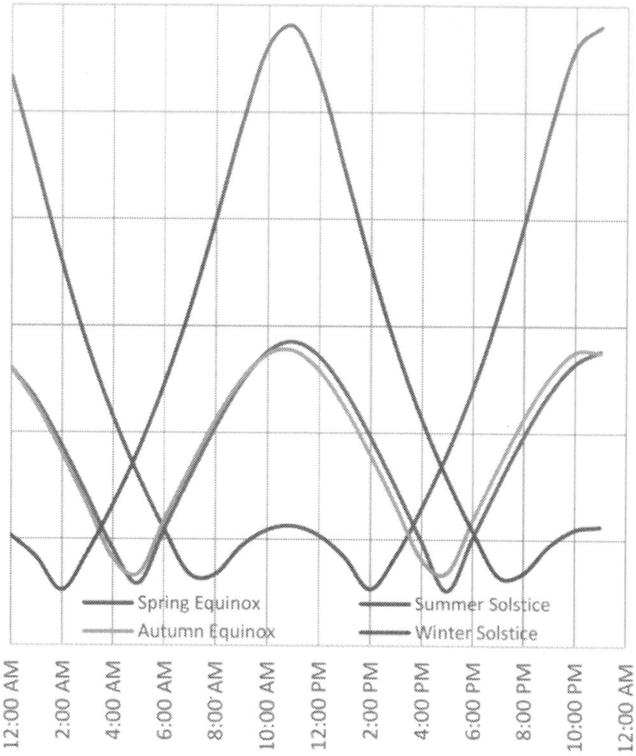

Figure 6. Daily variation of end loss at a specific location

The end loss is constantly changing depending on solar times and angles as mentioned above; however, the extent of that end loss depends on location, namely the latitude angle. The following polynomial curve fit was obtained by regressing end loss data at different latitudes

$$L_{End\ Loss}^{Maximum} = a + b\,\lambda + c\,\lambda^2 + d\,\lambda^3 + e\,\lambda^4 + f\,\lambda^5 + g\,\lambda^6 \quad (17)$$

Where a = *7.87873E-01*, b = *-9.18497E-05*, c = *2.75105E-03*, d = *1.70851E-07*, e = *-1.53553E-06*, f = *-6.15529E-11*, g = *4.91693E-10*

The polynomial expresses the maximum possible end loss as a function of the latitude angle. Figure 7 plots the previous polynomial and shows that end losses increase significantly with latitude in both hemispheres.

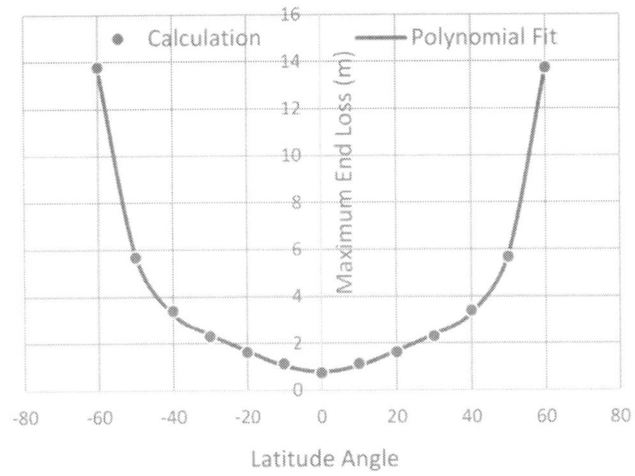

Figure 7. Global maximum end loss versus latitude

In addition, Figure 8 demonstrates the daily variation of end loss on an arbitrary day at different latitudes. The arbitrary day taken to show the daily variation was spring equinox selected for its average outcomes. The absolute value of the latitude angle was used to generalize results for both hemispheres. It is evident from Figure 8 that end loss increases with latitude as mentioned before.

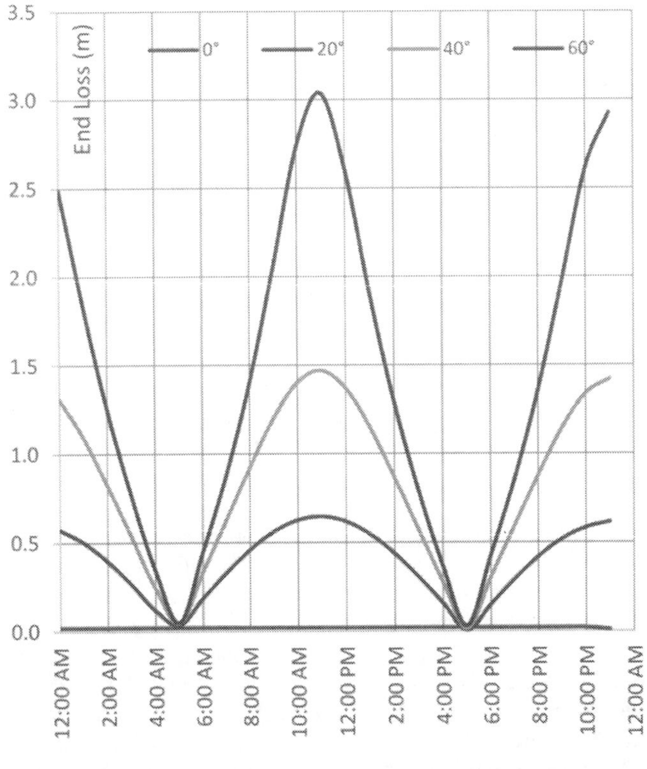

Figure 8. Spring equinox end loss trends at different */ λ /*

A performance model of a 50 MW_e net rated capacity PTC type CSP plant fitted with a TES system will be used to evaluate gains realized by implementing the new PTC design. CSP performance modeling has been described in detail before [1] so has TES operation and control [16]. Transient operation is the norm in CSP plants; therefore, previously described model modifications were adapted in this performance model to properly account for transient start up and shut down operations as well as cloud cover events and HTF residence time in the SF [17]. The model was run to simulate an annual operation at eight different locations around the world shown in Table 1. The model was not altered to optimize the size of the SF with respect to available DNI at each location. Thus, the eight different simulations differed in only four items: the three constants shown in Table 1 and DNI data.

Table 1: Selected CSP plant locations

Location	*λ (°)*	*l (°)*	*TZ (h)*
Adelaide, Australia	-34.93	138.52	9.50
Antofagasta, Chile	-23.43	-70.43	-4.00
Delingha, China	37.37	97.37	8.00
Aswan, Egypt	23.97	32.78	2.00
Bikaner, India	28.05	73.35	5.50
Casablanca, Morocco	33.37	-7.58	0.00
Sevilla, Spain	37.42	-5.90	1.00
Palm Springs, USA	33.83	-116.50	-8.00

The above equations were coded in *Microsoft Excel*, where the *GRG Nonlinear* engine of its *Solver* add–in was used to find the optimum PTC–absorber misalignment, *B*, that will maximize total annual DII at each location based on its TMY records: $\Sigma\ (DII \cdot \Delta t)_{Maximum} = f\ (B_{Optimum})$. Table 2 lists total DNI taken from the TMY weather files of NREL as well as the maximum reached end loss and optimum PTC–absorber misalignment found at each location.

Table 2: Optimum PTC–absorber misalignment found

Location	*Σ DNI (kWh/m²)*	$L_{End\ Loss}^{Maximum}$ *(m)*	*B (m)*
Adelaide, Australia	1,691	2.76	0.48
Antofagasta, Chile	2,044	1.82	0.37
Delingha, China	1,544	3.00	0.79
Aswan, Egypt	2,434	1.86	0.40
Bikaner, India	1,739	2.14	0.58
Casablanca, Morocco	1,583	2.58	0.48
Sevilla, Spain	1,773	3.03	0.57
Palm Springs, USA	2,534	2.65	0.50

DNI normally decline with latitude as can be inferred from TMY files or other empirical solar radiation models; nonetheless, clouds, humidity, dust, aerosols, and other uncontrolled natural phenomena can result in deviation from this general rule. Totalized DNI records taken from TMY files shown in Table 2 validate this principle. It is also evident that the maximum reached end loss and optimum PTC–absorber misalignment increase with latitude as can be seen in Table 2. It is imperative to restate that optimum PTC–absorber misalignment results listed in Table 2 were obtained based on TMY data which should be sufficient to show the benefits of the proposed misalignment; though, actual DNI and other weather conditions vary significantly from TMY data on everyday operation. Therefore, it may be a better practice to optimize the size of the proposed PTC–absorber misalignment based on DNI records from an empirical cloudless model rather than TMY data.

Next, the aforementioned performance model of a 50 MW_e net rated capacity PTC type CSP plant fitted with an adequately sized TES system will be run in three different PTC–absorber misalignment modes to assess the effect of the proposed new PTC design on solar heat collection, electric power generation, and plant revenues. The three PTC–absorber misalignment modes are: standard mode with no misalignment shown in Figure 3, idealized mode with a constantly moving HCE shown in Figure 4, and the recommended mode with the optimized PTC–absorber misalignment shown in Figure 5. Then, annual simulation results for the three different PTC–absorber misalignment modes at each location will be compared to assess the outcome of the proposed end loss attenuation scheme. At each location, results are presented in three categories: *Normal* refers to the standard mode with no misalignment producing the worst results, *Fixed* refers to the proposed PTC–absorber misalignment mode, and *Tracking* refers to the idealized mode with a constantly moving HCE in the lateral direction yielding the best possible results.

Table 3 shows the annual model output of solar heat forwarded to the PB for direct electric power generation. The amount of heat transfer slightly increased when the proposed PTC–absorber misalignment was implemented by 0.1 – 0.4%. The amount of heat transfer can increase by 0.2 – 1.0% if a continually tracking HCE is used. This slight increase is due to having an oversized SF designed to simultaneously meet the heat demand of the PB and thermally charge the TES system. Table 4 shows the annual model output of excess solar heat sent to the TES system for later reclamation. The amount of heat transfer modestly increased when the proposed PTC–absorber misalignment was used by 0.3 – 1.1%. The amount of heat transfer can increase by 0.9 – 2.4% if a continually tracking HCE is used. The above results will be different if different PTC dimensions were used.

Table 3: Heat transfer from SF to PB, GWh_{th}

Location	*Normal*	*Fixed*	*Tracking*
Adelaide, Australia	296.14	296.46	298.13
Antofagasta, Chile	391.16	391.37	391.92
Delingha, China	262.41	263.53	264.98
Aswan, Egypt	465.50	465.87	466.38
Bikaner, India	399.47	399.72	400.75
Casablanca, Morocco	294.48	294.69	295.48
Sevilla, Spain	331.07	331.38	332.91
Palm Springs, USA	458.44	459.15	459.69

Table 4: Heat transfer from SF to TES, GWh_{th}

Location	*Normal*	*Fixed*	*Tracking*
Adelaide, Australia	116.14	116.57	117.13
Antofagasta, Chile	178.00	178.55	179.64
Delingha, China	50.42	50.97	51.64
Aswan, Egypt	233.49	234.13	235.62
Bikaner, India	65.55	66.08	66.75
Casablanca, Morocco	97.02	97.56	98.15
Sevilla, Spain	106.36	106.95	107.66
Palm Springs, USA	194.55	195.79	196.70

If the results of Table 3 and Table 4 are combined, the amount of solar heat collected in the SF can increase by 0.1 – 0.5% if the proposed PTC–absorber misalignment is used and by 0.4 – 1.2% if a continually tracking HCE is used. With reference to location, least improvement was realized at Antofagasta, Chile while most improvement was realized at Delingha, China.

Table 5 presents the annual model output of electric power generation. The amount of electricity generated modestly increased when the proposed PTC–absorber misalignment was applied by 0.1 – 0.5%. The amount of electric power generation can increase by 0.4 – 1.2% if a continually tracking HCE is used. With regard to location, least growth was realized at Antofagasta, Chile while most growth was realized at Delingha, China.

Table 6 shows the annual model output of CSP plant revenues based on a 50 $/MWh_e$ net renewable electricity value. Revenues modestly increased when the proposed PTC–absorber misalignment was applied by 0.1 – 0.5% matching the above generation results. Revenues can increase by 0.4 – 1.2% if a continually tracking HCE is used matching the above generation results. With regard to location, least growth was realized at Antofagasta, Chile while most growth was realized at Delingha, China.

Table 5: Gross power generation, GWh_e

Location	Normal	Fixed	Tracking
Adelaide, Australia	153.18	153.45	154.28
Antofagasta, Chile	212.58	212.86	213.51
Delingha, China	114.07	114.66	115.39
Aswan, Egypt	262.28	262.64	263.41
Bikaner, India	171.04	171.36	172.00
Casablanca, Morocco	145.10	145.39	145.98
Sevilla, Spain	162.74	163.13	163.95
Palm Springs, USA	244.30	244.81	245.68

Table 6: CSP plant revenues, $

Location	Normal	Fixed	Tracking
Adelaide, Australia	$6.89 M	$6.91 M	$6.94 M
Antofagasta, Chile	$9.57 M	$9.58 M	$9.61 M
Delingha, China	$5.13 M	$5.16 M	$5.19 M
Aswan, Egypt	$11.8 M	$11.82 M	$11.85 M
Bikaner, India	$7.7 M	$7.71 M	$7.74 M
Casablanca, Morocco	$6.53 M	$6.54 M	$6.57 M
Sevilla, Spain	$7.32 M	$7.34 M	$7.38 M
Palm Springs, USA	$10.99 M	$11.02 M	$11.06 M

CONCLUSION

A modification of the standard PTC design intended to reduce HCE end losses is suggested. The modification consists of purposefully misaligning PTC from its HCE absorber in the axial direction by a preset amount. The structure of the PTC remains essentially the same in the newly proposed design; however, the HCE will no longer precisely align with the collector due to the deliberate but calculated axial misalignment. The cost of the new PTC is expected to be equal to that of a typical PTC since the only difference between the two PTC designs is a small axial PTC–absorber misalignment.

The size of the proposed misalignment was optimized to minimize HCE end losses based TMY weather data at eight international sites. *Solver*, an optimization engine available as a *Microsoft Excel* add–in, was employed to find the ideal misalignment magnitude at each location that will cause maximum incident solar radiation. A performance model of a PTC type CSP plant was then utilized to evaluate the outcome of the proposed axial PTC–absorber misalignment on solar heat collection, electric power generation, and plant revenues. The gains varied with location but were generally modest; even so, they will be likely attained at no extra cost.

The US Department of Energy launched the *SunShot* Initiative in early 2011 with the aim of reducing the cost of employing solar energy to levels equivalent to the cost of employing conventional energy by the end of the decade. The current exploration is another attempt in a series of attempts intended for increasing the viability of exploiting solar energy. More enhanced CSP modeling techniques were disclosed before [1] [17]; moreover, CSP operating procedures to manage HTF flow in SF [2] and control the charging and discharging processes of TES systems [16] were suggested to improve operation and upturn profits.

Also, a scheme to streamline the usually oscillating power output of CSP plants under transient conditions was proposed [18] and currently pending patent approval [19]. The proposed scheme slightly decouples solar heat supply from sink allowing for more control of heat transfer rates thus smoothing out electric power generation. The proposed scheme aims to solve the problem of electricity oscillations usually associated with the intermittent solar energy thus tolerating more contribution from CSP plants to the electric grid. Moreover, electric power generation slightly increased under the proposed scheme because the PB ran at a slightly higher thermal efficiency due to better dispersion of solar heat to the PB.

Doping SF HTF with nanoparticles was also explored [20] but was found to be practically useless since it had almost no effect on power generation not to mention the extra cost of nanoparticles and arising new operational challenges like sedimentation and abrasion. The current study falls within the same theme of reducing the cost of CSP deployment to make it more feasible as it aims to attenuate the negative impact of the inevitable absorber end loss. Meeting the target of the *SunShot* Initiative will most likely transpire by a number of minor incremental improvements rather than a single silver bullet.

Other potential prospects to optimize CSP operation to meet the target of the *SunShot* Initiative may involve altering the tilt and orientation angles of PTC assemblies. adjusting the tilt angle is very challenging as it will involve sloping the entire SF, but adjusting the orientation angle, polar rather than north–south positioned PTC axis, seems plausible. Pressurized HTFs currently used in PTC type CSP plants boil around 400 °C. If higher boiling HTF were to be developed, the PB working fluid can reach higher temperatures and subsequently raising thermal efficiency. Incorporating a TES system in CSP operation decouples heat supply from sink a bit thus raising its dispatchability. Operation Research tools can then be used to optimize CSP dispatch to maximize revenues based on a validated performance and financial plant model. Lastly, integrating other operations with CSP, like desalination and chemical processing, can synergistically increase feasibility.

NOMENCLATURE

a	azimuth angle, radians
AT	altitude transverse, radians
AW	aperture width, m
B	optimized misalignment
C	mirror cleanliness, percent
CSP	concentrating solar power
Day	day of year, day
DII	direct incident insolation, W/m²
DNI	direct normal insolation, W/m²
FL	focal length, m
HA	hour angle, radians
HCE	heat collection element
Hour	hour of day, hour
HTF	heat transfer fluid
HXT	heat exchanger train
IA	incident angle, radians
IAM	incident angle modifier, percent
L	length, m
l	longitude angle, degrees
NREL	National Renewable Energy Laboratory
PB	power block
PTC	parabolic trough collector
RD	row distance, m
SA	shadow argument
SD	solar day, day
SF	solar field
SH	solar hour, hour
TC	time correction term, minute
TES	thermal energy storage
TMY	typical metrological year
TZ	time zone, hour
α	altitude angle, radians
δ	declination angle, radians
Δx	PTC–absorber lateral misalignment
η	efficiency, percent
λ	latitude angle, degrees

REFERENCES

[1] M Abutayeh, DY Goswami, EK Stefanakos. Solar Thermal Power Plant Simulation. Environmental Progress & Sustainable Energy Journal. July 2013.

[2] M Abutayeh, A Alazzam, B Khasawneh. Balancing Heat Transfer Fluid Flow in Solar Fields. Solar Energy Journal. July 2014.

[3] S Al-Hanaei, S Al-Shomali, M Abutayeh. Performance Model of Shams I Solar Power Plant. Heat Transfer Engineering Journal. January 2016.

[4] Jie Sun, Ruilin Wang, Hui Hong, Qibin Liu, An optimized tracking strategy for small-scale double-axis parabolic trough collector, Applied Thermal Engineering, Volume 112, 2017, Pages 1408-1420, ISSN 1359-4311.

[5] Ruilin Wang, Wanjun Qu, Hui Hong, Jie Sun, Hongguang Jin, Experimental performance of 300 kWth prototype of parabolic trough collector with rotatable axis and irreversibility analysis, Energy, Volume 161, 2018, Pages 595-609, ISSN 0360-5442.

[6] Wanjun Qu, Ruilin Wang, Hui Hong, Jie Sun, Hongguang Jin, Test of a solar parabolic trough collector with rotatable axis tracking, Applied Energy, Volume 207, 2017, Pages 7-17, ISSN 0306-2619.

[7] Wanjun Qu, Ruilin Wang, Hui Hong, Jie Sun, Hongguang Jin, Prototype Testing of a 300kWth Solar Parabolic-trough Collector Using Rotatable Axis Tracking, Energy Procedia, Volume 105, 2017, Pages 780-786, ISSN 1876-6102.

[8] Fahim, U. and M. Kang "Performance Evaluation of Dual-axis Tracking System of Parabolic Trough Solar Collector." IOP Conference Series: Materials Science and Engineering 301(1): 012166.

[9] George C. Bakos, Design and construction of a two-axis Sun tracking system for parabolic trough collector (PTC) efficiency improvement, Renewable Energy, Volume 31, Issue 15, 2006, Pages 2411-2421, ISSN 0960-1481.

[10] Juan Fang, Qibin Liu, Taixiu Liu, Jing Lei, Hongguang Jin, Thermodynamic evaluation of a distributed energy system integrating a solar thermochemical process with a double-axis tracking parabolic trough collector, Applied Thermal Engineering, 2018, ISSN 1359-4311.

[11] A. E. Gaafar and A. F. Zobaa, "Economical design of a two-axis tracking system for solar collectors," 5th IET International Conference on Renewable Power Generation (RPG) 2016, London, 2016, pp. 1-6.

[12] Ming Li, Chengmu Xu, Xu Ji, Peng Zhang, Qiongfen Yu, A new study on the end loss effect for parabolic trough solar collectors, Energy, Volume 82, 2015, Pages 382-394, ISSN 0360-5442.

[13] Lippke F. Simulation of the Part Load Behavior of a 30MWe SEGS Plant. SAND95-1293. Sandia National Laboratories; Albuquerque, NM: 1995

[14] Cohen GE, Kearney DW, Kolb GJ. Final Report on the Operation and Maintenance Improvement Program for Concentrating Solar Power Plants.

SAND99-1290. Sandia National Laboratories; Livermore, CA: 1999.

[15] Rheinländer J, Bergmann S, Erbes MR. Technical and Economic Performance of Parabolic Trough Solar Power Plants – A Computational Tool for Plant Feasibility Studies. 14th SolarPACES International Symposium on Concentrated Solar Power and Chemical Energy Technologies; Las Vegas, NV: 2008.

[16] M Abutayeh, A Alazzam, B Khasawneh, Optimizing Thermal Energy Storage Operation, Solar Energy Journal, October 2015.

[17] M Abutayeh, A. Alazzam, Adapting Steady-State Solar Power Models to Include Transients. Solar Energy Engineering Journal, April 2017.

[18] M Abutayeh, A Alazzam, B Khasawneh, Streamlining the Power Generation Profile of Concentrating Solar Power Plants, ASME 12th International Conference on Energy Sustainability, Lake Buena Vista, FL, June 2018

[19] US Patent Pending: System And Method To Stabilize Power Generation Of Concentrated Solar Power Plants, Filed in July 2017.

[20] M Abutayeh, Y Addad, E Abu–Nada, A Alazzam, Doping Solar Field Heat Transfer Fluid with NanoParticles, Solar Energy Engineering Journal, February 2019.

CITATION

M Abutayeh, K Jeong A Alazzam, B El-Khasawneh, *Attenuation of Absorber End Loss in Parabolic Trough Collectors*, ASME 2019 13th International Conference on Energy Sustainability, Bellevue, WA, July 2019

574593 Modal Switching for Continuous Solar Reforming of Methane

The carbon footprint of methane reforming processes can be reduced by using solar thermal energy in place of fossil energy for process heating. In this scheme, concentrated solar power provides heat that drives endothermic reformation of methane, storing renewable energy in the chemical bonds of the resultant synthesis gas. CO_2 can be directly valorized in this process to further offset end-use carbon emissions. However, solar transience hinders the operational window and overall efficiency of these systems, limiting their practical feasibility. Here predictive switching between bi-reforming and tri-reforming of methane modes over a single nickel-based catalyst is applied in order to operate solar methane reforming continuously. This is performed using Oregon State University's solar simulator and directly-irradiated volumetric reactor, which is designed to promote fast transient behavior of solar-thermochemical reactions. Solar-to-chemical efficiency is calculated over periods of transient modal switching, where solar power is reduced incrementally and oxygen with additional methane are throttled into the feed to compensate for this decrease in external heating. Catalyst stability and coke formation are examined after modal switching.

567405 Sustainable Steam on Demand for the Bayer Process

Sustainable Steam on Demand for the Bayer Process

Alicia Bayon and Andrew Beath

CSIRO Energy, PO Box P.O. Box 330, Newcastle, NSW 2300, Australia

In this work, we explore the coupling of Concentrated Solar Thermal (CST) to produce the steam required for the Bayer Process used for the conversion of bauxite to aluminium hydroxide. Specifically, we present the use parabolic trough and central receiver tower technologies to produce the required steam for the digestion and electricity demand of a commercial-scale Bayer Process plant (392 MW_{th}). This work explores the integration of different configurations for the production of steam using annual simulations with real direct normal irradiation data. The effect of the solar field size as a critical parameter towards the economic optimisation of the plant is investigated. In addition, we analyzed plausible sites based on physical and operational constraints. We fixed two targets, a minimum CST capacity factor of 29% and a minimum NPV of $20 million. The internal rate of return (IRR) is also used to find the optimum design conditions at a specific location.

Results for trough systems are shown as an example at Learmonth. Other sites in Australia like Gladstone, Pinjarra and Darwin (Australia) are also explored. In this abstract, Learmonth is shown as an example of achieving the target CST input of minimum 29% at relatively small field and small storage sizes (see white line in Figure 1.a), and achieving positive net present value (NPV) results that increase as the system size increases (see background in Figure 1.a). Other sites like Gladstone also achieves good results, with a local NPV maximum at solar multiple 2 with 2 hours of storage providing a CST input of approximately 34%. Darwin, does not appear to be as promising for the technology and negative NPVs result from all combinations meeting the target CST input. Selection of the optimum at Learmonth is limited by the maximum IRR obtained which occurs at solar multiple 2 with 6 hours of storage. The IRR values are over 12% in most of the design conditions for Learmonth and Gladstone. The maximum is localized in solar multiple 2 and 6 hours storage for Learmonth where the IRR is 17.0% (Figure 1.b).

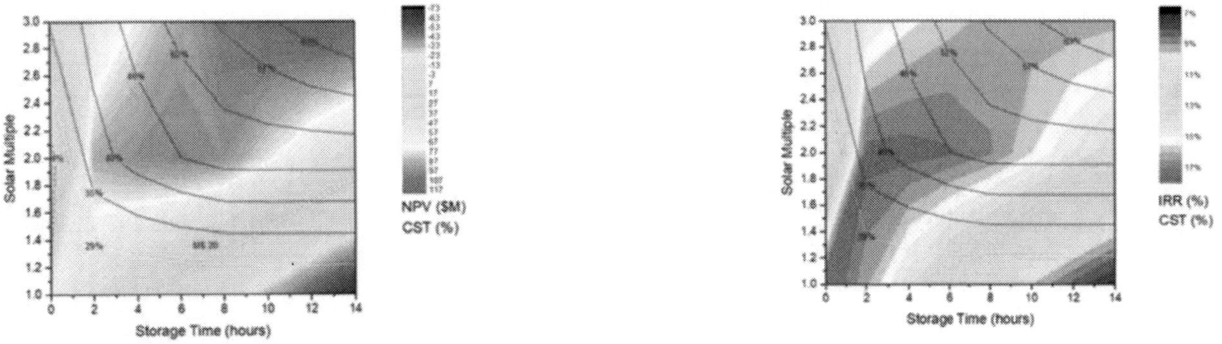

Figure 1: Net present value (NPV), internal rate of return (IRR) and capacity factor (CST) of parabolic trough plants (392 MW_{th}) using the cost model at Learmonth.

574517 Modal Switching for Continuous Solar Reforming

The carbon footprint of methane reforming processes can be reduced by using solar thermal energy in place of fossil energy for process heating. In this scheme, concentrated solar power provides heat that drives endothermic reformation of methane, storing renewable energy in the chemical bonds of the resultant synthesis gas. CO_2 can be directly valorized in this process to further offset end-use carbon emissions. However, solar transience hinders the operational window and overall efficiency of these systems, limiting their practical feasibility. Here predictive switching between bi-reforming and tri-reforming of methane modes over a single nickel-based catalyst is applied in order to operate solar methane reforming continuously. This is performed using Oregon State University's solar simulator and directly-irradiated volumetric reactor, which is designed to promote fast transient behavior of solar-thermochemical reactions. Solar-to-chemical efficiency is calculated over periods of transient modal switching, where solar power is reduced incrementally and oxygen with additional methane are throttled into the feed to compensate for this decrease in external heating. Catalyst stability and coke formation are examined after modal switching.

556448 Experimental Study on a Solar Falling-Film Photochemical Hybrid System for the Decolorization of Azo Dyes

Based on the solar photochemical reactors with flow channel (SPC), a new hybrid system that integrated solar falling-film photochemical reactor with photovoltaics (SFPC) was first presented to achieve the dye decolorization and the electricity generation simultaneously. Persulfate was chosen as the photochemical oxidant, all the experiments were performed outdoors in Tianjin, China. In this work, the spectrum transmittance, decolorization efficiency and the PV panel parameters were analyzed, by dealing with different concentration of acid red 26 (AR26) and acid brown 14 (AB14) in different systems.

The structural modification on the reactor improved the optical transmission efficiency of the SFPC system. The actual optical loss was only 1.6% in the SFPC system in contrast of 21.9% in the SPC system. The results confirmed that the remove of borosilicate glass and the reduced water thickness greatly decreased the optical loss in the SFPC system. The performance of the dye decolorization and the PV panel parameters were also tested. Compared with the SPC system, the SFPC system greatly enhanced the decolorization efficiency, the accumulated ultraviolet needed shortened 33.64%, the I_{sc}, P_{max} and of the PV panel increased 26.97%, 30.47%, 2.9% respectively, when dealing with the 50 mg/L AR26. And we will do the research on the degradation mechanism of the dyes in the future work.

The flow rate of the wastewater was also optimized. When the flow rate increased from 200 L/h to 1000 L/h, the reaction kinetic constants of AR26 and AB14 increased by 0.3 and 0.97 times respectively. However, the output power of the panel decreased and the energy consumption to maintain the operation of the system increased 4.93%. Taking the high energy loss with little rise of decolorization efficiency, 200 L/h is best choice for the system operation.

565598 Optimal Start-up Policies for a Nanofluid Based Solar Thermal Power Plant

Continuous increment in human population, along with the depletion of conventional fossil fuels and global warming, have propelled the development of renewable and sustainable technologies for energy production such as photovoltaics technologies and solar thermal plants, which allow recovering the solar energy. In this sense, parabolic trough solar power plants use a thermal fluid to transfer thermal energy from solar radiation to a Rankine cycle in order to drive a turbine that, coupled to an electrical generator, produces electricity. In order for these technologies to truly compete with conventional energy systems, research has been focused on the improvement of their performance either through closed-loop control of thermal solar plants or through the employment of different working fluids such as nanofluids, produced by the addition of nanoparticles to a base fluid. However, optimal start-up procedures of such plants, and an operational analysis when using diverse nanofluids (Al_2O_3-water and TiO_2-water) compared against the base fluid, are yet to be addressed.

Thus, this work seeks a highly sturdy approach for the optimal start-up procedure of such-mentioned plants, focused on power delivery and the time required for the system to reach target conditions. Proposed approach consists on energy and mass balances for all components of the system, constraints for flow through the collector and for maximum filling volume of storage tanks, as well as an evaluation of the power delivered by a coupled ORC when certain conditions are reached. Also, this work presents a comparison of various start-up scenarios when alternating between nanofluids and pure water as the working fluid on the collector. In addition, mathematical model was formulated as a Mixed Integer Nonlinear Programming problem where diverse case studies were solved off-line using GAMS software for their later implementation using feedback control systems.

Results show how optimal control of the working fluid volumetric flow aids the system to reach normal operation in a short period of time, maximizing the power delivered in the time range of study and preserving energy on call stored for the plant to keep producing energy afterwards. Regarding the working fluid, results illustrate that TiO_2-water nanofluid is the most suitable among the ones studied, since its heat transfer characteristics allow for a higher amount of energy recovery at the solar collector and its exchange at the boiler of the ORC, leading to both minimum start-up time and maximum power delivery, and in some cases, achieving more energy stored at the tanks. It is worth noting that saturation of the fluid might occur if optimal start-up policies are not tracked, bringing out deposition of nanoparticles and, consequently, damaging the equipment. Therefore, the importance of deploying these policies is key for an optimal and reliable operation, not only in early stages of operation, but also through the entire lifetime of the system.

567368 Two-Cell Photoelectrochemical Water Splitting Prototype Demonstration System

Hydrogen generation by photoelectrochemical (PEC) water splitting is a promising technology for solar energy storage. However, large scale application of PEC systems in their current single-cell design is technically challenging and economically questionable since the hydrogen is produced at the solar site. Accordingly, a PEC hydrogen production plant requires hermetic sealing of the entire solar array, construction and maintenance of an immense hydrogen piping infrastructure, and constant sensing for hazardous leaks. In our previous paper[1], published in 2017 in *Nature Materials*, we presented our solution to these issues by separating the hydrogen and oxygen cells using solid-state redox mediators based on nickel hydroxide. This allows for the oxygen-generating PEC cell to remain completely open, eliminating the need to hermetically seal the solar site, fit it with membranes and collect the hydrogen from that site. The hydrogen is then generated in a separate, compact electrochemical cell directly at the end-user's location. Thus far, a few PEC cell designs have been proposed based on the single-cell configuration. In this work, we present a first of its kind prototype of a two-cell PEC device based on a 100 cm^2 hematite photoanode stack and commercial Si PV modules. Battery-grade nickel hydroxide electrodes are used as redox mediators between the cells. This prototype device has been designed and optimized for continuous PEC hydrogen production. The issues of material selection, stability, light harvesting and operation scheme are addressed and discussed.

[1] Landman, A., Dotan, H., Shter. G. E., Wullenkord, M., Houaijia, A., Maljusch, A., Grader, G., and Rotschild, A. (2017). *Photoelectrochemical water splitting in separate oxygen and hydrogen cells.* Nature Materials, March, 1-7.

573226 Iron-Nickel Alloy Films for Neutral Electrolyte Electrochemical Reactions

Increasing global energy demand and environmental concerns have pushed towards the development of alternative energy conversion and storage systems. Hydrogen is viewed as an ideal alternative fuel source which can be produced through the coupling of renewable energy (e.g., wind, solar) to processes such as electrochemically-driven water electrolysis. The majority of commercial water electrolyzers currently operate in acidic or alkaline environment. However, the most abundant source of water is seawater. Traditional catalysts require seawater to be desalinated, an energy-intensive process that increases the cost for hydrogen production. Thus, hydrogen production in neutral conditions presents its self as a viable substitute. However, electrochemical hydrogen production in neutral electrolyte suffers from slow kinetics compared to alkaline and acidic electrolyte.

This work is centered on the development of (FeNi) alloy films for their application as electrocatalysts for hydrogen production in a neutral electrolyte. Herein, we present the synthesis, characterization, and electrocatalytic evaluation of (FeNi) alloy films. Alloy films were synthesized through electrochemical deposition. Further energy-dispersive x-ray spectroscopy (EDX), inductively coupled plasma – mass spectroscopy (ICP-MS), and x-ray photoelectron spectroscopy (XPS) were used to determine the chemical composition and surface chemistry of the as-synthesized films. Cyclic voltammetry (CV) and chronoamperometry (CA) were performed on the (FeNi) alloy films to evaluated their catalytic performance for hydrogen production. Through electrochemical evaluation, the optimum candidates for hydrogen evolution in a neutral environment were identified.

557923 Fresh Look on Wind Power Generation-Upgraded, 25/02/2019

Fresh look on wind power generation-Upgraded, 25/02/2019

Author: David Judbarovski, systems engineering, retired engineer, Akko, Israel

judbarovski@gmail.com

First version and publication of that breakthrough technology goes back to July, 2017 and was titled as "Fresh look on wind power generation". That was really a fresh & new look now upgraded into a system, can produce by 0.25 US cent/kWh (because a lion share of the cost is gearless permanent magnet electric generator of USD 80.0/kW) and deliver that electricity reliably and safely to any point of the world in practically unlimited quantity by final cost less than 1.0 USA cents per kWh for any consumption schedule.

The key idea originally was to transform a route energy of a sail boat into rotation of a generator equipped by a hydraulic turbine with effectiveness near 100%.

So the power generation unit is a small pilotless floating structure (SPFS) with a simplest flat sail and relatively small hydraulic turbine is fixed underwater to the SPFS moved by the sail.

The said SPFS go there and back. They harvest the wind energy and transform it into electricity.

Certainly, we can ask me a way to store and to transport such electricity from the thing floating in the open ocean?

We can transform such electricity in hydrogen-gas thermally from desalinated sea water, then collected on-board for further transportation to floating terminals.

The hydrogen-gas can be liquefied at the terminals and transported in thermal insulated vessel by air at high altitude by hydrogen blimp (i.e. airship being a simplest inflatable balloon filled by hydrogen at small pressure), using a negligible share of liquid hydrogen as the said blimp fueling. The liquefying will increase a cost of end electricity for consumers vs. its sea/ocean origin one, but totally it is much more cheaper, quicker, safer and directly than by tankers, if more than 150 ton blimp for about 10,000 km and then for electricity producing for micro-grids by hydrogen-air fuel cells. It is suitable for long-term storage or for transforming it in other energy carriers, e.g. in ammonia or hydrocarbons.

Some examples of tremendous consequences of my invention in economics and human life:

(1) Such cheap and limitless and easy accessible electricity will turn a human life, production and transportation to electricity usage as practically only energy resource and transform design of things and tools for energy effective feeding by electricity.

(2) In such case, there would be enough about 40.0 trillion kWh electricity annually in a middle-term perspective to feed all needs of mankind. It can be provided by 4.0 milliards kW power can be provided by 160.0 million simplest SPFS-s of 25 kW each, much simpler and in much less quantity than cars now used, and by total CAPEX of about USD 250 milliards annually during 5 years. Really, it is practically negligible expense in comparison with world GDP. We can imagine a way to solve the all world energy supply problem practically free? Here I did it, and did by green way. I can think that in our turbulent world there are much more reliably to prefer electricity generation in far open ocean than in a nearest sea, even if the electricity would be twice more expensive.

577341 A Dynamic Genome-Scale Metabolic Network Model for a Novel Methanotroph-Cyanobacteria Coculture

A Dynamic Genome-scale Metabolic Network Model for a Novel Methanotroph-Cyanobacteria Coculture

Kiumars Badr, Q. Peter He and Jin Wang*

Department of Chemical Engineering, Auburn University, Auburn, AL, 36849, USA. *Email: wang@auburn.edu

Biogas is comprised primarily of methane (50%~70%) and carbon dioxide (30% ~50%), which can be produced from various waste sources, including landfill material, animal manure; wastewater; and industrial, institutional, and commercial organic wastes. EPA estimates that currently US biogas production potential is 654 billion cubic feet per year, which could displace 7.5 billiongallon of gasoline [1]. It is clear that biogas has immense potential as a renewable feedstock for producing high-density fuels and commodity chemicals. However, the utilization of biogas represents a significant challenge due to its low pressure and presence of contaminants such as H_2S, ammonia, and volatile organic carbon compounds. To tap into this immense potential, effective biotechnologies that can co-utilize both CO_2 and CH_4 are needed.

In our previous work, we have demonstrated that metabolic coupling of methane oxidation to oxygenic photosynthesis can be a highly efficient way to recover the energy and capture carbon from biogas [1, 2]. However, development of multi-organism platforms for commercial biogas conversion present significant challenges which center around our ability to control function and composition of species in the Coculture. An essential tool for the optimization, design and analysis of the coculture based biogas conversion is the development and validation of kinetics models that can accurately describe and predict the co-culture growth under different conditions [3]. To this end, using *Methylomicrobium buryatense - Arthrosipira platensis* as the model coculture system, we have developed an unstructured model to capture the growth dynamics. Specifically, Monod-like models were developed to capture coculture growth. Two sources of substrate were considered in the model: gas transferred from gas phase and gas produced *in situ*. In addition, we rely on the fitted maximum cell growth rate for both strains to capture other potential interactions. Using designed experiments and the developed model, we clearly demonstrated that the synergistic effect within the coculture cannot be fully explained by the *in situ* substrate exchange, and there must be other "metabolic links" to explain the significantly enhanced cell growth of both strains in the coculture.

In order to test different hypothesized "metabolic links" between the Methanotroph and cyanobacteria in the Coculture, in the work, we developed a structured, dynamic genome-scale metabolic network model for the Coculture system. These GEMs, especially the dynamic GEMs, offer a comprehensive picture of cellular metabolism and serve as a bridge that can better link the work of microbiologists and engineers in understanding and optimizing complex cellular metabolism. By integrating the available knowledge on each strain with data obtained in our own experiments, we use DFBAlab to implement the dynamic GEM for the Coculture [4]. For the dynamic GEM, besides the GEMs for each individual strain, the key inputs to the DFBA model are the uptake kinetics for different substrates. In this work, the substrate update rates are provided from the unstructured dynamic model we already developed. In addition, the product secretion rates and cell growth rates predicted by the validated unstructured model were used to validate the coculture dynamic GEM. Although the experimental measurements to validate the model are mainly limited to cell growth, substrate consumption and product secretion rates for each organism in the coculture, the overall dynamic trajectory (both measured low-frequency samples and unstructured model predicted high-frequency samples) offers significantly more power in validating the model, compared to a few steady state conditions. The validated coculture GEM enables the testing of different hypotheses on potential "metabolic links" within the coculture.

Reference:

[1] Badr, K., Hilliard, M., He, Q.P., & Wang, J., "Understanding the Stability and Robustness of a Methanotroph-Cyanobacterium Coculture through Kinetic Modeling and Experimental Validation." Annual AIChE Meeting. Pittsburgh, PA (2018).

[2] Badr K, Roberts N, Hilliard M, He QP, Wang J. "Photoautotroph-Methanotroph Coculture – A Flexible Platform for Efficient Biological CO_2 – CH_4 Co-utilization," Dynamics and Control of Process Systems, including Biosystems - 12th DYCOPS. (2019); accepted.

[3] Almquist J., M. Cvijovic, V. Hatzimanikatis, J. Nielsen, M. Jirstrand, "Kinetic models in industrial biotechnology – improving cell factory performance", Metab. Eng., 24 (2014), pp. 38-60.

[4] Gomez JA, Höffner K, Barton PI. DFBAlab: a fast and reliable MATLAB code for dynamic flux balance analysis. BMC Bioinformatics. (2014),15:409.

567780 Kinetics and Reactor Design for Pyrolytic Treatment of Petroleum-Contaminated Soils

Scope and Significance

While marine oil spills from offshore platforms or tankers attract most of the public attention, the vast majority of oil spills (about 98%) occur on land. Every year about 10-25 million gallons of petroleum products are spilled mostly from pipelines and fixed facilities. Petroleum hydrocarbons pose long term threats to groundwater quality, inhibit the germination and root elongation rates of plants and are extremely toxic to important soil microbes. Moreover, polycyclic aromatic hydrocarbons (PAH) present in petroleum can damage the immune systems of aquatic organisms and wildlife.

Current remediation methods for petroleum-contaminated soils are either relatively slow or have unintended consequences in the form of soil damage and high-energy usage. Even worse, some processes such as aerobic bioremediation can activate toxic hydrocarbons and transform them to more noxious byproducts such as PAH derivatives. For all these reasons, there is still a pressing need for more efficient and *sustainable remediation* of petroleum-contaminated soils.

Pyrolysis is receiving increasing attention as an on-site remediation approach because of its potential to rapidly and reliably remove total petroleum hydrocarbons (TPH) with lower energy requirements and better post-treatment soil fertility than other *ex situ* thermal remediation approaches [1-3]. Pyrolytic treatment of contaminated soils at 420°C with only 15-min residence time in a continuous reactor reduced TPH by 99.9%, lowered the total PAH concentration by almost 95%, and restored soil fertility to 98% of the clean soil level [3]. These results clearly suggest that pyrolysis has the potential for improved ecosystem restoration following remediation compared to traditional thermal technologies. However, the same study also revealed potential tradeoffs between pyrolytic treatment intensity, soil detoxification efficacy and soil fertility restoration. Specifically, treatment at 470°C for 15- or 30-min reduced soil fertility to 51% and 39% of the clean soil level, which was only marginally higher that the fertility of the contaminated soil.

The existence of an optimal treatment intensity for contaminated soils underscores the need for mechanistic insight into the pyrolysis process and for a deeper understanding of how the detoxification of contaminated soil and the restoration of soil fertility are affected by (a) the properties of both the soil and the petroleum crude, and (b) the operating conditions of pyrolysis reactors.

Kinetics and Reactor Design for Pyrolytic Treatment of Petroleum-Contaminated Soils

Taking the first steps towards a rational design of pyrolysis reactors, we have combined mathematical techniques and advanced analytical measurements to:

1. Develop a full kinetic model for the pyrolytic treatment of soils contaminated with petroleum hydrocarbons. The new model describes all thermally-induced transformations of soil components (e.g. clays, carbonates, or SOM), the evaporation of light hydrocarbons and the pyrolysis of the heavier hydrocarbon fractions.
2. Develop a mathematical model describing the operation of continuous reactors (rotary kilns) that will be used for *ex situ* treatment of petroleum-contaminated soils. This dynamic model will then be used to estimate the optimal (a) pyrolysis temperature, (b) reactor residence time, and (c) gas flow rates that will achieve a specified TPH reduction at the minimum energy cost for any given throughput of contaminated soil.

Two clean soils with different compositions were spiked with oil crudes to prepare petroleum-contaminated soils. Thermogravimetry with evolved gas analysis (TG-IR and TG-MS) was used first to identify the mineral transformations taking place as the clean soils were heated in an anoxic atmosphere. The same analytical techniques were applied to study the two distinct processes occurring during the pyrolytic remediation of contaminated soils. Desorption of the lighter hydrocarbons is the dominant process for temperatures between 150 and 350°C. Thermal cracking reactions dominate in the 350-500°C range releasing pyrolysis products (like hydrogen, methane, or olefins) and leading to the formation of a carbonaceous solid material (char) that remains in the treated soil [2]. It is important to note that hydrocarbon desorption and pyrolysis overlap with the temperature ranges over which soil mineral transformations take place.

To develop our kinetic model, we assumed that the soil consists not only of inert components but also of "reacting pseudo-components" (e.g. clays, carbonates, SOM) that release either water or carbon dioxide as the soil is heated. Weight losses due to water or carbon dioxide release accounts for more than 95% of the weight loss observed during heating of clean soils. Contaminated soils contained additional pseudo-components that account for the weight losses observed due to the desorption of light hydrocarbons and the pyrolysis of the heavy petroleum fractions to yield gaseous products and char (coke). Using the distributed activation energy approach (DAE) approach, each pseudo-component was assumed to consist of a large number of species that react with activation energies distributed according to Gaussian probability density functions (PDF). To obtain the kinetic parameters, an optimization problem was formulated by fitting model predictions to experimental thermogravimetric data. Solving this nonlinear constrained optimization problem with an interior-point algorithm, the expected activation energies and variances of all pseudo-components were estimated, together with the preexponential factors for all the reactions and the initial mass fractions of all pseudo-components.

Excellent fits to the experimental data were obtained using four reacting pseudo-components for the clean soils. Three pseudo-components were required to model clay dehydration and dehydroxylation processes that occur over a wide range of temperatures and generate different multi-peak water release patterns for each soil. One pseudo-component was enough to model carbon dioxide release from either carbonate decomposition reactions or soil organic matter (SOM) pyrolysis. Two additional pseudo-components were required to model hydrocarbon desorption and pyrolysis.

The kinetic models were then used to simulate the operation of non-isothermal rotary kiln reactors that will be used for *ex situ* treatment of petroleum-contaminated soils. To preserve anoxic conditions, the reactors were heated externally while a stream of nitrogen gas flowed through the rotary kiln to sweep the desorbing hydrocarbons and pyrolysis products. A finite stage or mixing cell model with separate solid and gas phases was used to describe the operation of the pyrolysis reactor.

Model predictions were compared to experimental data we obtained with a continuous rotary kiln reactor [3]. The TPH reductions predicted by the model agreed very well with the experimental data for all the conditions considered in our earlier study: pyrolysis temperatures, 370, 420 and 470°C; residence times 15, 30 and 60 min; solid flow rates 7, 14, and 25 lb/h.

More importantly, however, our model was able to explain the trade-offs observed in our earlier study between pyrolysis treatment intensity, soil detoxification and soil fertility restoration [3]. The clay components of the soil used in [3] undergo significant dehydration between 350 and 450°C, a range that overlaps with that of pyrolysis reactions. As contaminated soil moves through the reactor zone kept at 420°C, the reaction rate for the heavy hydrocarbon fraction is fast enough to achieve almost complete conversion within 15 or 30 min. But, the same conditions are not severe enough to completely remove the water from the clay components of the soil. If the soil is treated at 470°C, on the other hand, the clay components are almost completely dehydrated, causing irreversible soil damage and loss of fertility.

By emphasizing the importance of soil composition and by raising the possibility of trade-offs between detoxification and fertility of treated soils, this study has significantly advanced our fundamental understanding of the *multiple* and *concurrent*physicochemical processes taking place during the pyrolytic treatment of petroleum-contaminated soils. Theoretical kinetic and reactor models are essential for designing pyrolysis reactors optimized for specific petroleum-soil systems, enabling robust operation and facilitating acceptance of this novel technology by regulatory agencies and stakeholders.

References

1. Vidonish, J. E.; Zygourakis, K.; Masiello, C. A.; Gao, X.; Mathieu, J.; Alvarez, P. J., *Environmental Science & Technology* **2016,** *50*, (5), 2498-2506.
2. Vidonish, J. E.; Alvarez, P. J. J.; Zygourakis, K., *Industrial & Engineering Chemistry Research* **2018,** *57*, (10), 3489-3500.
3. Song, W.; Vidonish, J. E.; Kamath, R.; Yu, P. F.; Chu, C.; Moorthy, B.; Gao, B. Y.; Zygourakis, K.; Alvarez, P. J. J., *Environmental Science & Technology* **2019,** *53*, (4), 2045-2053.

573966 Well Placement Optimisation in a Heterogeneous Petroleum Reservoir with Geological Uncertainty

ABSTRACT

Huge investments are required in the development of oil and gas fields; this implies that decisions at any stage of the development process must be aided by sound and proven mathematical-oriented methods, because the use of intuitive engineering judgement alone cannot guarantee sustainable profitability over long periods [1]. This presents an ever-increasing motivation to apply optimisation techniques for the enhancement of petroleum production from both planning and operational/scheduling perspectives. The planning aspect typically involves activities such determining the number and locations of wells, drilling the wells and infrastructure installation. On the other hand, the operational/scheduling aspect comprises of well rate allocation, flow scheduling and routing problems [2]. Incorporating optimisation at the planning stage is a pre-emptive step that mitigates subsequent operational difficulties over the field's lifetime. However, embedding optimisation in a planification problem can be more challenging compared to production and injection scheduling optimisation in an oil and gas field. Although many studies [3, 4] have addressed the operational/scheduling aspect of the field development problem, this work is premised on the proposition that, jointly optimising the planning and scheduling problem better enhances field operability.

The uncertainty of subsurface geology is another constraint to take into consideration when optimising field productivity [5]. To capture the uncertainties in the subsurface geological (reservoir) model, geostatistical realisations of the most accurate reservoir representation are obtained using available information (permeabilities, porosities, fluid saturation) [6]. These uncertainties in the model are often reflected in the diverse outcomes of well productivities and ultimately the field's Net Present Value (the objective function) [7, 8]. Given the numerous number of possible realisations, it is impractical to consider all realisations in the optimisation; hence, it is important to apply classical reduction methods to ensure computational feasibility.

In this work, we specifically address the well placement planning and rate control optimisation problem under geological uncertainty. The difficulty of this problem is characterised by the presence of discrete variables, nonlinear and nonconvex objective function and constraints. Increased complexity arises from the computationally demanding functional evaluations from full-scale reservoir simulation, which is done in ECLIPSE®. We also exploit the MATLAB® Reservoir Simulation Toolbox – MRST® [9] for interfacing with ECLIPSE® and performing derivative-free optimisation tasks. The developed workflow is applied to a synthetic case study, for which it is demonstrated using several operational scenarios that using optimally selected realisations reduces computational time while providing best operating well locations. A comparative analysis of the performance of different derivative-free optimisation algorithms is also presented.

REFERENCES

1. Epelle, E.I. and Gerogiorgis, D.I., Optimal Rate Allocation for Production and Injection Wells in an Oil and Gas Field for Enhanced Profitability. *AIChE J.*, **in press** (DOI: 10.1002/aic.16592).
2. Tavallali, M.S. et al., 2014. Well placement, infrastructure design, facility allocation, and production planning in multireservoir oil fields with surface facility networks. *Ind. Eng. Chem. Res.*, 53(27), 11033-11049.
3. Gunnerud, V., Conn, A. and Foss, B., 2013. Embedding structural information in simulation-based optimization. Comput. *Chem. Eng.*, 53, 35-43.
4. Kosmidis, V.D., Perkins, J.D. and Pistikopoulos, E.N., 2004. Optimization of well oil rate allocations in petroleum fields. *Ind. Eng. Chem. Res.*, 43(14), 3513-3527
5. Li, L., Jafarpour, B. and Mohammad-Khaninezhad, M.R., 2013. A simultaneous perturbation stochastic approximation algorithm for coupled well placement and control optimization under geologic uncertainty. *Comput. Geosci.*, 17(1), 167-188.

6. Wang, H., Echeverría-Ciaurri, D., Durlofsky, L. and Cominelli, A., 2012. Optimal well placement under uncertainty using a retrospective optimization framework. *SPE J.*, 17(01), 112-121.
7. Bellout, M.C. et al., 2012. Joint optimization of oil well placement and controls. *Comput. Geosci.*, 16(4), 1061-1079.
8. Gerogiorgis, D.I., Georgiadis, M., Bowen, G., Pantelides, C.C. and Pistikopoulos, E.N., 2006. Dynamic oil and gas production optimization via explicit reservoir simulation. *Comput. Aided Chem. Eng.*21, 179-184.
9. Lie, K.A., 2014. An introduction to reservoir simulation using MATLAB: user guide for the Matlab Reservoir Simulation Toolbox (MRST). *Sintef ICT*, Trondheim, Norway.

569460 Modeling of Thermal Cracking of Natural Gas Liquids: Optimization with Support Vector Machine-Based Constraint Handling Scheme for Stiff ODEs

Rapid increase in shale gas production in the Appalachian Basin (Marcellus and Utica Shales) for the past 5 years has led to a significant growth in natural gas liquids (NGLs) production as well as a projected increase of these petrochemical feedstocks in the upcoming years [1, 2]. In this perspective, many existing ethylene crackers have expanded capacity and new crackers are becoming online to benefit from this unique opportunity [3]. Hence, the mathematical modeling and optimization of this process emerges as a necessity so as to determine the optimal operating conditions and the reactor configurations for the steam cracker, in such a way that the profit from thermal cracking of NGLs is maximized.

In this work, we are demonstrating (i) the comprehensive first-principles modeling of an adiabatic propane steam cracking process as a 1D plug flow reactor with coking effects [4] and (ii) the data-driven optimization of this reactor model using the ARGONAUT [5-6] framework. The mathematical model for thermal cracking includes the reaction rates expressed with Arrhenius equation, as well as the ODEs for continuity, energy, and momentum balances. The kinetics of thermal cracking of propane is adapted from Sundaram and Froment (1979) and the cracking model is validated using industrial data [7]. This simulation model is later used in the data-driven optimization phase to collect input-output data. However, the stiffness in the reactor models poses an additional complexity in retrieving the optimal operating conditions for the cracking process, especially when the data-driven optimization strategies strongly rely on collecting numerically stable samples from the reactor simulator. To this end, a novel Support Vector Machine (SVM)-based filtering approach is applied to eliminate candidate sampling points that are susceptible to the stiffness in the model and that will result with an abrupt termination of the cracking simulation due to high pressure drop. The SVM classifier identifies and removes the numerically unstable candidate points *a priori* to simulation execution by formulating an optimal separating hyperplane between stable/unstable points. This enables the data-driven algorithm to consistently converge to a profitable and feasible reactor configuration for the propane cracking process with an improved objective function value.

References

[1] Energy Information Administration. Today in Energy: Appalachia, Permian, Haynesville drive U.S. natural gas production growth, August 2018. URL https://www.eia.gov/todayinenergy/detail.php?id=36934. (accessed 04.10.2019).

[2] Energy Information Administration. Short-term energy outlook, April 2019. URL https://www.eia.gov/outlooks/steo/. (accessed 04.10.2019).

[3] Energy Information Administration. U.S. ethane consumption, exports to increase as new petrochemical plants come online, February 2018. URL https://www.eia.gov/todayinenergy/detail.php?id=35012. (accessed 02.26.2018).

[4] O. Onel. (2017). Advances in Modeling, Synthesis, And Global Optimization of Hybrid Energy Systems Toward the Production of Liquid Fuels and Olefins. Princeton University.

[5] F. Boukouvala and C. A. Floudas. (2017). ARGONAUT: AlgoRithms for Global Optimization of coNstrAined grey-box compUTational problems. *Optimization Letters*, 11(5), 895-913.

[6] B. Beykal, F. Boukouvala, C. A. Floudas, N. Sorek, H. Zalavadia, E. Gildin. (2018). Global optimization of grey-box computational systems using surrogate functions and application to highly constrained oil-field operations. *Computers & Chemical Engineering*, 114, 99-110.

[7] K. M. Sundaram and G. F. Froment. (1979). Kinetics of coke deposition in the thermal cracking of propane. *Chemical Engineering Science*, 34, 635-644.

574887 Towards a Novel Energy Price Predictive Framework: The Texas A & M Energy Price Index

The energy markets are sensitive and volatile to technological breakthroughs, changes in monetary policy, or major global challenges. Moreover, various governmental agencies, political and commercial organizations, think tanks as well as researchers and academics around the globe, consider various energy policies and their effects in an attempt to deal with the increasing concerns in energy independence, energy scarcity, energy sustainability, and pollution caused by the utilization of energy. Also, in the current environment in which all strategic political and commercial decisions and policies are assessed in economic terms, it is of outmost importance to determine the cost of energy precisely so as to evaluate their effectiveness. Considering the fact that energy affects every single individual and entity in the world, it is crucial to quantify "the unit price of energy" accurately, and study how it evolves with respect to time, major breakthroughs, challenges, energy and monetary policies. Given the lack of such a tool, we develop a novel framework, the Texas A&M Energy Price Index (EPI) that can be used as a benchmark to calculate the average price of energy to the end-use consumers in the United States.

The complicated energy landscape of the United States is carefully depicted to determine the products that are directed to the end-use sectors of the U.S. economy [1-5]. The total energy demand of these products, together with their prices, serve as the backbone of the Energy Price Index [6-10]. Since several key components (e.g. prices, demands) of the Energy Price Index have a lag between two and three months, we introduce a rolling horizon based parameter estimation model to approximate the current value of the Energy Price Index. The predictive methodology is tested on 120 months to show the accuracy of the developed framework in the prediction of the current value of the Energy Price Index. Energy price sub-indices for the transportation, residential, commercial, and industrial sectors are also developed. Some of the potential applications of the proposed index in the areas of economics, finance, engineering, law, and policy are demonstrated.

References:

- Elia, J. A.; Baliban, R. C.; Xiao, X.; Floudas, C. A. Optimal energy supply network determination and life cycle analysis for hybrid coal, biomass and natural gas to liquid (CBGTL) plants using carbon-based hydrogen production. Computers and Chemical Engineering 2011, 35, 1399-1430.
- Elia, J. A.; Baliban, R. C.; Floudas, C. A. Nationwide energy supply chain analysis for hybrid feedstock processes with significant CO2 emissions reduction. AIChE Journal 2012, 58 (7), 2142-2154.
- Elia, J. A.; Floudas, C. A. Energy Supply Chain Optimization of Hybrid Feedstock Processes: A Review. Annual Review of Chemical and Biomolecular Engineering 2014, 5 (1), 147-179.
- Onel, O.; Niziolek, A. M.; Floudas, C. A. Integrated biomass and fossil fuel systems towards the production of fuels and chemicals: state of the art approaches and future challenges. Current Opinion in Chemical Engineering 2015, 9, 66-74.
- Floudas, C.A.; Niziolek, A. M.; Onel, O.; Matthews, L.R. Multi-Scale Systems Engineering for Energy and the Environment: Challenges and Opportunities. AIChE Journal 2016, 62 (3), 602-623.
- Energy Information Administration (EIA), Annual Energy Outlook, 2019
- Energy Information Administration (EIA), Monthly Energy Review, 2019
- Lazard, Levelized Cost of Energy Analysis Version 12.0, 2018
- Bureau of Labor Statistics (BLS), Producers Price Index, 2018

U.S. Department of Energy (DOE), Alternative Fuels Data Center, 2018

575666 Intensification of Biorefinery Operations Using Building Blocks

Bioenergy is an attractive form of energy since it requires renewable sources that can be produced locally and used as raw materials for its production. One of the possible routes for bioenergy production is through lignocellulosic biorefineries, which utilize various biomass sources, e.g. corn stover, wood, rice straw and forest wastes, to produce biodiesel, steam and power in addition to pulp and paper [1]. Increasing awareness towards sustainable and modular technologies highlights the need for novel processing pathways leveraging on process intensification principles that can yield drastic improvements in several metrics, including economics, energy consumption and chemical footprint [2]. Recent developments in process intensification resulted in development of Integrated biorefinery concept with two new processing pathways: (i) pre-extraction of hemicellulose from biomass and its conversion to ethanol, and (ii) production of syngas from gasification of black liquor [3]. However, there exist further unexplored opportunities for intensification of the biorefineries to enhance the reaction yield and mass/heat transfer within the several parts of the process, e.g. digestion of pulp, evaporation of black liquor, etc. [4].

Recently, a new method for systematic process intensification, which relies on building block-based representation of chemical process flowsheets, has been introduced [5]. These building blocks can be integrated to form a two-dimensional grid embedding numerous non-trivial process alternatives [6-9]. Each block can be assigned with several physicochemical phenomena, e.g. reaction, vapor-Liquid equilibrium, solid-liquid equilibrium, etc., combination of which can yield various intensification pathways. In this work, we extend this novel representation methodology to various operations that take place in a biorefinery. Pulping which occurs in a pulp digester lies at the heart of a pulp mill and, here, we develop a new representation based on fundamental building blocks to represent the pulp digestion process. Mass transfer boundary layer is represented by the boundary shared by two neighboring blocks in different phases and digestion process is represented within the interior of the block. This building-block based representation is modeled using a mixed-integer nonlinear programming (MINLP) problem so as to minimize the energy consumption and increasing the yield of the process while searching for several intensified pathways. The proposed approach is demonstrated through several case studies focusing on the conventional Kraft pulping process, Kraft pulping process with pre-extraction of hemicellulose, Kraft pulping process with pre-extraction of hemicellulose and short fibers.

[1] Huang, H., Lin, W., Ramaswamy, S., Tschirner, U. (2009) Process Modeling of Comprehensive Integrated Forest Biorefinery – An Integrated Approach. *Applied Biochemistry and Biotechnology, 154*, pp. 205-216.

[2] Tian, Yuhe., Demirel, S.E.., Hasan, M.M.F., Pistikopoulos, E.N., (2018) An overview of process systems engineering approaches for process intensification: State of the art. *Chemical Engineering and Processing - Process Intensification, 133*, pp. 160-210.

[3] Aden, A., Ruth, M., Ibsen, K., Jechura, J., Neeves, K., Sheehan, J., Wallace, B., Montague, L., Slayton, A., Lukas, J., (2002) NREL Report TP-510-32438. Lignocellulosic Biomass to Ethanol Process Design and Economics Utilizing Co-Current Dilute Acid Prehydrolysis and Enzymatic Hydrolysis for Corn Stover.

[4] Huang, H., Ramaswamy, S., Al-Dajani, W.W., Tschirner, U. (2010) Process modeling and analysis of pulp-mill based integrated biorefinery with hemicelluloses pre-extraction for ethanol production: A comparative study. *Bioresource Technology 101*, pp. 624-631.

[5] Demirel, S.E., Li, J., Hasan, M.M.F., (2017) Systematic process intensification using building blocks. *Computers and Chemical Engineering, 105*, pp. 2-38.

[6] Li, J., Demirel, S. E., Hasan, M. M. F., (2018) Process Synthesis Using Block Superstructure with Automated Flowsheet Generation and Optimization. *AIChE Journal, 64(8)*, pp. 3082-3100.

[7] Li J., Demirel S.E., Hasan M.M.F. (2018) Process Integration using Block Superstructure. *Industrial &*

Engineering Chemistry Research, *57(12)*, pp. 4377–4398.

[8] Li, J., Demirel, S.E., & Hasan, M. M. F. (2019). Building Block-Based Synthesis and Intensification of Work-Heat Exchanger Networks (WHENS). *Processes, 7(1)*, 23.

[9] Demirel, S. E., Li, J., and Hasan, M. M. F. (2019). A General Framework for Process Synthesis, Integration, and Intensification. *Industrial & Engineering Chemistry Research.* DOI: 10.1021/acs.iecr.8b05961.

576096 Hyperbolicity of Heat Transport Processes

Heat transport models are ubiquitously present in chemical, petrochemical, manufacturing and process industry as mathematical models of distributed parameter systems given by parabolic partial differential equations (PDEs). The conductive (diffusive) transport over macroscopic length scale is characterized by the well known Fourier's law and a large number of research efforts have been realized in the realm of modelling, control and optimization of such models. However, the salient characteristic of these models is that any initial disturbance in the material body is propagated instantly due to the parabolic nature of the partial differential equation (PDE) model. To eliminate this unphysical feature, we revisit the Fourier law to take into account the thermal inertia, resulting in a hyperbolic heat equation which avoids the phenomenon of infinite speed of propagation of initial data [1,2].

The main conservation laws are embedded in a modelling variety provided by the hyperbolicity of the transport systems which is physically relevant and desired property as action at distance is precluded and physically meaningful finite speed of phenomena propagation is ensured. Furthermore, the hyperbolicity mathematically ensures the well-posedness of local Cauchy problems [3]. In addition, a Stefan problem is considered, which is a specific type of heat transport boundary value problem for a partial differential equation describing the heat distribution evolution in a phase changing medium [4]. Since the moving interface is unknown a priori, the solution needs to take into account determination of the moving boundary position as well as accurate heat propagation. Finally, we provide derivation of the hyperbolic heat transport equations, the stability analysis and the numerical results for simple heat diffusion problem on fixed domain and Stefan problem.

Literature Cited

1. Cattaneo C. On a form of heat equation which eliminates the paradox of instantaneous propagation. *C R Acad Sci Paris*. 1958;pp. 431–433.

2. Vernotte P. Les paradoxes de la theorie continue de l'equation de la chaleur. *C R Acad Sci Paris*. 1958;246:3154–3155.

3. Fischer AE, Marsden JE. The Einstein evolution equations as a first-order quasi-linear symmetric hyperbolic system, I. *Communications in Mathematical Physics*. 1972;28(1):1–38.

4. Gupta S. *The Classical Stefan Problem. Basic Concepts, Modelling and Analysis*. North-Holland: Applied mathematics and Mechanics. 2003.

569568 Constructing Efficient Surrogate Models for High-Dimensional First Principles Systems Under Uncertainty

A plethora of chemical engineering processes are informed by sophisticated numerical simulations based on first-principles models (e.g., consisting of a set of coupled partial differential equations) that exhibit highly nonlinear behavior and are computationally expensive. It is often the case that the parameters in these models as well as the initial and boundary conditions in dynamical systems are uncertain and, thus, this uncertainty is propagated to the model outputs. Therefore, relying on deterministic simulation results could possibly lead to erroneous outcomes for the design, operation, and control of these processes. The computational burden associated with solving the entire set of equations can be prohibitive when one wishes to run the model for the whole spectrum of possible input parameter values, even in this era of high-performance computing, especially in the context of optimization applications. Metamodeling techniques aim at alleviating this problem by approximating first-principles model responses under uncertainty with efficient black-box mappings that can serve as surrogate models for the original high-fidelity models [1]. The primary notion in metamodeling is to calibrate the surrogate models using the fewest possible runs of the original model without compromising their predictive capability.

In this work, we explore the effectiveness of various well-established meta-modeling techniques applied to a basic, but relevant crystallization problem, i.e., the growth/dissolution of a 2D crystal population through cooling/heating cycles. This technique is used, possibly in combination with milling, to manipulate the crystal shape of active pharmaceutical ingredients, and efforts have been recently devoted to optimize and control this process. Even though relatively simple, such a problem represents an interesting test-case for metamodeling since it is highly nonlinear, inherently dynamic, and industrially relevant. [2,3] The crystallization system is described by a system of integro-partial differential equations, whose parameters are inherently uncertain. The parameter uncertainties are modeled by suitable probability distributions selected according to available experimental data and expert knowledge.

We first employ sparse polynomial chaos expansions (PCE) whose coefficients are estimated based on least-angle regression and explore their scalability with an increasing number of uncertain inputs. PCEs have been traditionally used for uncertainty quantification in numerous applications (e.g., see [4]). We then present the fundamentals of Gaussian Processes (i.e., Kriging) [5] and explore their coupling with sparse PCEs in order to create surrogate models with optimal local and global predictive capabilities [6]. Finally, the use of NARX neural networks [7] is demonstrated to deal with the continuous dynamic behavior of the system. The accuracy of these three surrogate modeling approaches in capturing the highly nonlinear spatio-temporal dynamics of the crystallization process at hand is assessed using state-of-the-art cross-validation and validation measures, and their computational aspects are discussed.

[1] Bhosekar A., Ierapetritou M., "Advances in surrogate based modeling, feasibility analysis, and optimization: A review", Computers and Chemical Engineering, 108:250-267, 2018

[2] Salvatori, F., Mazzotti, M., "Manipulation of particle morphology by crystallization, milling, and heating cycles—A mathematical modeling approach", Industrial and Engineering Chemistry Research, 56:9188–9201, 2017.

[3] Bötschi, S., Rajagopalan, A. K., Morari, M., Mazzotti, M., "Feedback control for the size and shape evolution of needle-like crystals in suspension. I. Concepts and Simulation Studies", Crystal Growth and Design, 18:4470–4483, 2018.

[4] Paulson J.A., Mesbah A., "An efficient method for stochastic optimal control with joint chance constraints for nonlinear systems", International Journal of Robust and Nonlinear Control, 1-21, 2017.

[5] Kawai S., Shimoyama K. "Kriging-model-based uncertainty quantification in computational fluid dynamics. 32nd AIAA Applied Aerodynamics Conference, 2014, 2737.

[6] Schöbi R., Sudret, B., Wiart J., "Polynomial-Chaos-based Kriging", International Journal for

Uncertainty Quantification, Begell House Publishers, 5:171-193, 2015.

[7] Mai C.V., Spiridonakos M.D., Chatzi E.N., Sudret B., "Surrogate modeling for stochastic dynamical systems by combining nonlinear autoregressive with exogenous input models and polynomial chaos expansions", International Journal for Uncertainty Quantification, 6:313-339, 2016.

AUTHOR INDEX